How to Lose Your Farm in Ten Easy Lessons and Cope with It

How to Lose Your Farm in Ten Easy Lessons and Cope with It

Robert Hitt Neill
and
James R. Baugh

Mississippi River Publishing Co.
Leland, Mississippi

Printed in the United States

Cover artwork by Bob Tompkins
Designed by JoAnne Prichard

The authors wish to thank the Oklahoma Conference of Churches for permission to quote from *AG-Link Report,* December 4, 1985, by M. L. Brock.

MISSISSIPPI RIVER PUBLISHING CO.
P.O. Box 882
Leland, MS 38756

Library of Congress Cataloging-in-Publication Data

Neill, Robert Hitt.
How to lose your farm in ten easy lessons and cope with it.

1. Family farms--Mississippi River--Delta--Finance.
2. Bankruptcy--Psychological aspects. I. Baugh, James R.
II. Title.
HD1476.U6A156 1987 338.1'3'097624 87-14176
ISBN 0-9617591-1-9

Contents

Prologue

He had awakened from a troubled sleep at 4:30 A.M., and had given up and put the coffee water on to boil. Outside the cabin, it was drizzling rain, perfectly normal for the first day of deer season, he thought as he stirred down the coffee for the last time. Gazing out at the darkness, alone, he sipped the black brew. His brother had quit and moved away the last winter, disheartened by having to harvest in the mud and lose money again farming. He stared out the window and reflected, "No light a'tall out there—just like there's no light at the end of my tunnel."

The farm had been in the family for three generations, but now he was the last one left and he could see no hope that there would be another generation on the land. He had farmed since returning from the service but didn't see how he could last much longer. "What's the use of keeping on?" he muttered as he refilled the cup. Outside, bare branches became dimly visible in the greying, weeping east.

All yesterday and last night, his thoughts had not been on his hunting, but on his family and the farm. Broken-hearted, he had finally admitted that the only future for them was in his life insurance. He had weighed every alternative carefully, and it came down to the fact that with the insurance money, his wife could pay off the farm debt, still have some left over, and rent out the place for enough to provide a decent living for herself and the two children. He could think of no way they would be better off with him than without him.

He stepped over to the gun rack where his .30-06 deer rifle hung. It was a three-gun rack with a row of pegs across the bottom. Setting his coffee cup down, he took the rifle and checked the chamber to be sure it was empty. He then slipped the trigger guard over one of the pegs and propped the barrel in the lower rack. When he stood in the doorway to the bedroom, the barrel rested only eight inches from his chest. He clicked the safety off, took a deep breath and pulled gently on the barrel.

There was a loud "click" as the firing pin fell on the empty chamber.

He took the gun from the rack, worked the pump action to recock it, and positioned it as before. Again, a gentle pull on the barrel produced the "click." He tried it twice more and, satisfied, replaced the rifle in the rack and poured himself more hot coffee. He stood at the window and sipped it in deep meditation. He heard a couple of jeeps crank up at the other cabins, but his mind was not on deer hunting.

Even though the coffee was steaming hot, he felt cold down deep in his bones. He stifled a shiver in the muscles of his shoulders and chest. When the cup was empty, he set it on the back of the sofa and walked slowly into the bedroom to dress.

He dressed carefully, even putting on fresh underwear. He started to stuff the day-old underwear and shirt into his duffle bag, but changed his mind and threw it carelessly on the unmade bed. "Doesn't need to look too neat;" he thought, "needs to look like an accident." With that in mind, he even donned rubber boots, rain pants, and parka. After visiting the bathroom one more time, he walked slowly around the den, gazing long at the snapshots of happier times framed on the walls. He tested the trigger-pulling arrangement one more time before slipping a loaded clip into the rifle. Taking a deep breath, he pumped a bullet into the chamber, set the gun in the rack with the peg through the trigger-guard, and pushed the safety to "off." Standing sideways in the doorway, he took a deep breath and gripped the barrel.

A jeep with a loud muffler came roaring by the cabin in first gear, its occupants talking loudly over the noisy exhaust. He stepped back quickly into the shadows of the bedroom, lest his friends see him and hail him through the front door that stood ajar. They were out of sight down the road when the trembling overtook him, and he collapsed onto a bunk. His legs shook violently and his teeth chattered. The coldness went deep into his vitals.

"Heck, even at eight inches, I'd probably miss like

this," he muttered. He walked by the gun rack and poured another cup of coffee, the last in a six-cup pot. The hot brew did little to warm him, but the shaking did subside. Finally, purposely, he put the cup down, walked back to the gun rack, and positioned himself in the doorway. He knew where the heart was; he had seen much death in his time.

For some reason, he decided to check the arrangement one more time. He went through the ritual of removing the gun, unloading it, placing it back on the peg, pulling on the barrel until the firing pin clicked, removing the rifle, loading it, placing it on the rack, and knocking the safety off again. He stood in front of the Remington, his favorite, and wondered if anybody would ever use it again after he was found. "I hope so," he thought, and smiled sadly as he reached with his right hand to caress the crack in the oil-finished forearm.

"God, forgive me," he breathed as he shut his eyes and tugged the gun barrel toward his chest.

The peg broke.

He opened his eyes when the explosion didn't occur, and saw the peg lying on the sofa. Sighing, he snapped the safety back on. He picked up the peg and looked at it wearily. 'Hell, I can't even do that right," he muttered.

Shoulders drooping, he shuffled back to the stove and put the water on to boil for another pot of coffee.

He didn't go to the woods that morning. As he sipped

Book One

How to Lose Your Farm in Ten Easy Lessons

still another cup of coffee at the window, he thought, "Maybe God *is* weeping today. And maybe if I can write the story and sell it, something good will come of this. And maybe some other poor farmer will read it and recognize where he's going before he gets there."

R.H.N.

Once upon a time, the American Dream could be defined in two words: Own Land. Heckfire, in any cowboy picture show worth the price of the popcorn, you would find a variation of the following scene:

The long cattle drive is finally finished, the steers are in the corrals by the railroad station, and the boss is paying off the cowhands. The first two are Ace and Lucky. Lucky says to Ace, "What you gonna do wit' chores?"

Now, Ace is a lanky, saturnine man with a dark complexion, a handlebar moustache, and a low-slung holster containing a Colt with clearly visible notches in the grip. His eyes seem to smolder, and every viewer over puberty can tell he's still pining over a Secret Lost Love. Ace pockets his gold, rolls a cigarette with one hand, and gestures with the match toward a saloon down the street. "Reckon Ah'll see kin Ah find me a little caird game," he drawls.

At this point, you know that Ace will hold a great deal of whisky well, get caught dealing off the bottom of the deck, and have to shoot his way out of town sometime around dawn. He will not wing people; he coolly bores them dead center. As he rides off into the sunrise, why do you have the vague impression he is headed for banking school? Anyway, Ace is typical, but he is not your All-American Boy.

Lucky shoves his ten-gallon hat back to reveal carefree blue eyes and blonde curls. He jiggles his bag of gold up and down like he knows full well it will burn a hole in his pocket if he doesn't get rid of it fast. He spies a beautiful innocent-looking maiden who is smiling at him from a balcony over the saloon. Lucky jumps in the air and kicks his heels; "Yahoo!" he bellows, "Ah'm gonna find me a bottle and a gal and paint this town red!"

Actually, you know that Lucky is not really honing to paint anything, but is just expressing his youthful exuberance for wine, women, and song. He will have to be dunked in the water trough in the morning to be able to sit his horse, and will have successfully blown three month's pay in one night. You get the feeling that it was almost worth it, and you somehow know that Lucky will soon mature and marry some rancher's daughter. If he doesn't get in a card game with Ace.

Lucky is also typical, but he is not the All-American Boy. The All-American Boy is Curly, next in line to get paid off. As Ace saunters and Lucky gallops toward the saloon, the grizzled, sun-wrinkled, white-haired foreman turns to the clean-cut young man leaning against the corral fence. "Whut yew gone do, Curly?"

Curly blushes. "Aw shucks, Cap'n," (all ranch foremen are former Texas Rangers) "you know Ah bin spookin' that new schoolmarm. Me an' Sairy Jane bin lookin' at a li'l ole spread down toward Tombstone, an Ah b'lieve this'll 'bout make a down payment."

There you have it. The rest is automatic. The picture show may or may not end before the happily-ever-after part, but the dye is cast. Curly will lope off into the sunset atop his white stallion, marry the pretty schoolmarm, and buy the little spread out from Tombstone (whose owner was looking at a bigger spread down toward Flagstaff). The young couple will have four kids, fight off two Indian attacks, enlist the aid of a John Wayne-like gunfighter in a dispute with a neighboring cattle baron over water rights, and eventually Curly himself will become a cattle baron who lets everyone water at his streams.

So went the American Dream. If you had to be a doctor, lawyer, or Indian chief, you only practiced that occupation long enough to accumulate a down payment on that li'l ole spread down toward Tombstone.

Then, about fifty years or so ago, them li'l ole spreads got to be about as scarce as hen's teeth. Or at least, things got to where it took more than a couple of cattle drives to scrape together the down payment. In Curly's day, folks could still go west for bigger and greener pastures but then California was discovered, and the All-American Boys quickly realized that *nothing* was farther out than California. There ensued a couple-decade scramble to grab up the spots in between that no one had wanted earlier, but after that, land that would grow anything other than rocks was about full-up. Kids began to go to college, get law degrees, and practice professions like this for their whole lives.

Then politics was invented to give surplus lawyers something to do, and the country *sho' 'nuff* got in trouble!

Well, boys and girls, I am not particularly happy to announce it, but the day has come when once more the American Dream can flourish! Land, that commodity they ain't making no more of, is again becoming readily available. Those who had about become resigned to being lawyers and doctors and insurance agents for the rest of their lives can now latch on to The Dream. They are welcome to latch on to mine, if they so choose.

Now, I bet that a lot of folks picked up this little book saying, "What a cute title! Wonder what it's *really* about?" If you got this far, you no doubt think you are in a Zane Grey reprint. Let me put your mind at ease. The title is really what it says: I am going to tell you how to lose the farm. The "easy" part is easy to misunderstand, though. What I mean is that, for instance, it's "easy" to step off a ten story building, "easy" to fall, and "easy" to stop at the bottom. "Easy" gets used up fast, though, if you decide to change your mind between step and stop. So it is with farming.

I had in mind to tell you how to lose a Mississippi Delta farm, for the South is where my experience is.

Western farms, for that matter, are not even called farms, but ranches. And when a Westerner says it nowadays, it seems closely akin to something you tighten bolts and nuts with. I had a tractor driver that used to tighten his tractor battery cables with a pair of "plars an' a 'justable ranch." After he had been on one tractor for a month, you couldn't get the cables off with anything less than a hacksaw, and when it quit cranking because of corroded terminals, you just said the hell with it and pulled it off with a chain every morning. As far as I know, the easiest way to lose a "ranch" is to haul off and "chunk" it. I used to do that with Cliff's once a summer.

Someone once told me they had farms up north. I was always skeptical of this, since I was once in Philadelphia and it must have snowed three feet in November. My brother, Beau, who is normally a truthful man, was stationed in Newport, Rhode Island, while he was in the Navy, and he said there were farms up there. Said they grew stuff like rutabagas, which undoubtedly are similar to the carpetbagas that I used to hear my grandmother cuss about. If so, I wouldn't touch one on a dare.

The reason I don't believe there are really farms up north is that it is so cold up there! I have seen pictures of snow and ice that would be hard to believe had I not been there once and besides that, have you ever heard a Yankee talk?

When I was in the Navy, I was on a carrier that served as flagship for the whole fleet. I was in charge of getting on the radio and telling as many as forty or fifty other ships that the Admiral wanted us to turn to starboard and slow to ten knots, all at the same time and in formation, like the band at the Sugar Bowl halftime show.

They couldn't understand me.

Well, the long and the short of it is that the Navy, in its infinite wisdom, deemed it necessary to send me to school for the best part of a month to teach me how to talk to Yankees. When I got home two years later, the tenants on the place could not understand a word I said. I told Bootjack three times to go chisel plow the Dan and Doe cut, and he finally turned to my father and said, "Mist' Robert, whut do Mista Bob want me t'do an' whayr do he want me t'do it?"

It seems to me that the reason Yankees talk like that is because it is so cold up there that they're afraid their words will freeze, so they clip them off and spit them out quick. I got the impression that if a normal person would go to New York (assume for the sake of illustration that a normal person might visit New York) and drawl out "parabolic sub-soiler," that the words would freeze and stick out through his lips like a popsicle stick until he managed to get back below the Mason-Dixon line. I cannot believe that crops grow under those climatic conditions, so I admit to being totally unqualified to instruct you on losing a Yankee farm.

Unless you get the wrong impression, I ain't got anything in particular against Yankees. I just don't care about them winning the pennant unless someone from Ole Miss or State is playing with them.

I also understand that there are farms, though maybe no more farmers, in the Midwest. I feel sorry for these folks for several reasons. One is, I went to see a couple of picture shows about farmers in that area that made me want to see the manager and not only get my admission price back, but also the price of the popcorn. If those shows are true, those folks have already been going through what us Southerners are now catching—hell! I also pity them because they ain't got names. A Southern farmer used to be referred to as a "Planter"; a Western farmer, ignorant of nuts and bolts, prefers "Rancher," and Yankee farmers, if you believe the rumors, are probably "Carpetbaggas." But Midwestern farmers? You got me.

Last, I feel sorry for the Midwestern farmers because of their mistaken identity. Lots of these folks come South every year selling tools ("ranches"), or painting barns, or custom grain harvesting. My Daddy used to call them "Mule Gypsies," because in the old days they sold mules. Anyway, they were all nice people, but talked funny. Before I knew better, I asked one if he was a Yankee. His horrified reaction reassured me, but Daddy whipped me, and Mother washed my mouth out with soap anyway. I apologized and the fellow said he forgave me, but can you imagine going through life being mistaken for a Yankee?

Anyhow, with all the other problems these Midwesterners have, I ain't presuming to tell them how to lose their farms. If the movies were true, they could have told me how to lose mine a little quicker than I managed it on my own. But then, I don't know if I could have appreciated their advice.

In trying to give you the particulars of Southern farm losing, I am going to try to keep it light. I hope I can, not only for your sake, but for mine. Every now and then, I catch myself getting a little bitter and sarcastic, and I want you to know I didn't used to be like that. I learned it farming.

Let me illustrate that. I have a good friend who is a Ph.D. in Weed Science, or some such thing that regular farmers have to know without all the initials. Holland needed some test plots, so he rented out thirty acres from me, bought a used tractor and some machinery, and planted his milo and cotton. He had just rowed up the rest to plant in beans when the rain started. Five weeks later, he knocked on my door one morning early. "Come look at the field with me," he invited.

We walked across the grown-up rows, and he asked, "Is it dry enough to re-hip?"

"I believe it would be cloddy, if you did," I said."What about running a cultivator in the middles and double harrowing the tops of the row?" He thought that might work.

A while later, he was back at my house to borrow my jumper cables. His tractor, parked for five weeks, wouldn't start. An hour of useless cranking brought him back to ask me to take my tractor and pull his tractor off. I did so, and we finally got him running. Within another hour, he hunted me up to borrow one of my cultivators; his had broken. After we had disconnected his and re-connected mine to his tractor, he opined that he would fill up with diesel while he was there.

He had not known to leave the drum tilted.

Five weeks worth of rain collecting on the top of a barrel of diesel with loose bungs usually results in several inches of water in the drum. Holland's pump sucked the water off the bottom first. Soon he was knocking on the door again, asking to use my tractor to run the cultivator while his assistant drained his tractor's gas tank and replaced the fuel filters. By the time his tractor was running again, it was after lunch, and he decided to hitch to the harrow. He finished harrowing the first time right behind my tractor with the cultivator, so he switched my tractor to the harrow and hooked his up to the planter. Just as he filled the first hopper with seed, the clouds opened up again.

Holland barely had time to empty the hopper back into the sack and park both tractors, and by the time he got his truck out to the blacktop, he was skidding from side to side of the slick turnrow. I sat there in my truck, grinning, in case he didn't make it out. To me, it was another typical farmer's day. Not to him.

Fuming, he pulled up next to me and rolled his window down a crack. "Dad-gum Rain!!!" he complained. "I ain't ever gonna get these damn beans planted!"

Holland is a very religious person, whose father is a

preacher. I raised my eyebrows at him; "I don't think I ever heard you cuss before you started farming."

He seemed momentarily startled, thought for another moment, then replied wonderingly, "You know, I don't think I ever did!"

I went home and composed the following poem.

The Ph.D. and the Working Farmer

Holland Jordan came to town, typical preacher's kid.
He didn't smoke, he didn't drink, nor mess with those who did.
A cussword never passed his lips, not even in distress
And women never tempted him; his Lisa was the best!

Then Dr. Jordan bought a farm, it was out at Geneill;
For tractors and equipment, he then made another deal.
The weather for him held no fears, he had a Ph.D.
The weeds and bugs would run when that diploma they would see.

Farmer Jordan had a tractor, EE-I-EE-I-OH!
With water in the diesel fuel, that tractor would not go.
He showed it his diploma and the battery went dead,
And words that he had heard Bob say went dancing through his head.

His fields were overrun with grass, he had to hire a plane;
Degrees he used to threaten thrips all seemed to be in vain.
He waved with his diploma at the morning glory vines;

They seemed to laugh, and with his plants continued
to entwine.

Farmer Jordan's used equipment rusted up and broke;
When he began to talk to it, his words went up in
smoke!
The rains began, with muddy fields, his beans he
could not plant.
A drink he'd have to have 'fore bed, else he would rave
and rant!

He learned he has to cuss machines to make them
want to run.
The weather he cannot control, his Ph.D.'s no fun.
The chemicals he used on weeds just make them
want to grow;
So, "What the hell, double the rate, or triple it, like
Beau!"

The church will no more see his face, you'll find him
in a bar.
With Lisa and the kids at home, he gambles from afar.
He'll cuss blue smoke; poor Lisa's scared that he will
one day harm her.
For Dr. Holland Jordan has become a Working
Farmer!!

LESSON 1
The Buy-in

Most of the folks who are farming today . . . well, let's back up six or eight years, because that's about when all The Trouble started. And if you got a head start on the rest of us and lost your farm early, I don't want to exclude you from my readership. Especially if the wounds have had time to heal and you are looking at land prices and thinking about going back into the ring. For God's sake, read this first.

The vast majority of the folks who were in farming six or eight years ago had one thing in common: their land had been in the family for generations. Only a small minority, however, had gotten their land free and clear. For one reason or another, each generation usually had to buy in, and as land prices increased, so did mortgages. There were several reasons for this, and I'm not saying any of them are necessarily bad. All I'm doing is stating facts.

As folks had fairly large families up until this last generation (were *you* married before The Pill?), the farming father was faced with the choice of leaving some land to each child, or leaving the land to one or two kids and whatever else he had accumulated going to the rest. Generally, since the farmer had worked for the land all his life, he valued the farm more than such mundane things as money or bank stock. Certainly the kids would pick up on this attitude. What happened was this: one, the farm got split up so small that no one could support a family off it; or two, one or two of the kids had to buy out the others, usually at market prices, and this invariably involved mortgage financing; or three, the kids got to squabbling about it and then the lawyers got into it, and that was all she wrote.

Another kicker that you have to throw in here is inheritance taxes. I don't have any actual statistics on this, but I'd bet that many a farm has had to be sold for inheritance taxes, or mortgaged for enough that it was later lost or is in danger of being lost because of inheritance taxes. Many farmers skimp for their whole lives to pay off their land for their children, yet never make the necessary estate plans to keep it from having to be heavily mortgaged within a year of death.

I understand this. The last generation of farmers was simply not going to die. They were tough, independent men who just were not atuned to thinking about a negative aspect such as dying. My own father had not even made out a will until he had a near fatal aneurysm four years before his death. I believe I can report, however, that this kind of thinking has turned completely around. The present crop of farmers, if they've been lucky enough to be able to pay their life insurance premiums, walks around with lightning rods during thunderstorms, secure in the knowledge that their families will be better off if God's lightning strikes before the banker's.

There's an old saw that two things are certain: death and taxes. A good tax lawyer or CPA may not be able to help you avoid death, but he can sure help avoid some of the taxes if you don't wait until after you're dead to get around to seeing him. If you want to open a real can of worms, try leaving a five hundred-acre farm without a mortgage to three children under forty. A few years ago, the IRS would have said it was worth two or three thousand dollars an acre, for a total valuation of well over a million dollars, and with a straight face the man might have said you owed a couple hundred thousand in taxes. Then just as that mortgage gets in place, the fourth child, who had married a lawyer and who had been left fifty thousand in insurance money, says, "But, Daddy left you over three hundred thousand apiece!"

Few non-farmers ever seem to understand that farm land is worth one thing when you want to sell it, and something entirely different when you want to farm it.

For the sake of illustration, and without running on about it, let's look at a typical case:

Jimmy Ray is currently involved in losing his farm, which was founded by his great-grandpa. Here's its history. The thousand acres were cleared by Great-Grandpa Joe, who bought the land for fifty dollars an acre. He left the land to all four kids. Sally married Gus and separated her 250. Billy and Bobby bought out Susie's for $100 an acre. They each have three children.

Billy leaves his to Jim and Joe, and Joe later marries, has two children and dies in a car wreck leaving no will. Bobby leaves his to Sam and Jane, who marries Mike.

Jim buys back 175 acres at $300 an acre from two of Sally and Gus's kids. He later buys out Sam for $400 an acre. Jim has three children—Jimmy Ray, Ann, and Willy. Jimmy Ray buys his Uncle Joe's land from the widow and two children, the youngest of which has finally reached majority. He pays $500 an acre. After Jim's death, J.R. and Willy buy Ann's portion for $750 an acre. Ten years later Willy says to hell with it and leaves for Florida, deeding his land and mortgages to J.R.

Since they have had to roll over a mortgage payment or so after a couple of disaster years, Jimmy Ray now has 740 acres with about four hundred thousand owed on it. Five years ago he was offered seventeen hundred an acre for it, but turned it down, not from greed, but because he liked to farm. The land might now bring $750 an acre, but there would be an additional capital gains tax of over $150 an acre and what with delinquent interest and a realtor's fee, J.R. can no longer come out even with a sale. This is known as being between a rock and a hard place. Or better yet, remember the discussion on the use of "easy" in the first chapter.

What's the solution, you say? Well, I didn't start out promising answers. I am dealing exclusively in problems and do not intend to be sidetracked from my original mission. I am merely making the point that there are few fields which offer a young person the opportunity of becoming so deeply in debt so early in life as the family farm does.

Of course, you may be saying here that nobody holds a gun on a boy to make him go into farming. True, to a certain extent. But at the very least, it used to be expected, especially of an older son. Sample conversation around the coffee shop thirty years ago:

"You hear young Billy Benson is finishing school and going into architecture?"

"Well, I thought old Will had been dodging folks lately!"

"What the hell's wrong with the kid?"

This attitude may be changing, though. Just yesterday I heard down at the parts place:

"You hear young Freddie Fodrod's coming home from the Navy to help his Daddy farm?"

"Dumb little son-of-a-gun!"

Lesson 2
Traditional Philosophies

I had been home from the Navy less than six months and was getting broken in running the cotton gin. (Since you can't hear in a cotton gin anyway, it didn't make any difference that no one could understand how I talked.) I had taken some gin tickets over to the office and was relaxing with an R-O-C Cola when one of the customers came in. As he picked up his tickets, he was steady "po-mouthin'" about his crop. Finally, Uncle Sam told him, "Hush, man. You're makin' a bale and three quarters while everybody else is making a bale and a quarter!"

"Yeah," the customer said, "but it takes me two bales to break even!"

After he left, my uncle looked long and hard at me and said, "Son, remember that. If you can farm close to the vest and make a little money off a bale to the acre, you're better off than breakin' even at two bales."

As in any profession, there are two basic philosophies in farming. You can try to hold costs down and make a little money on a small volume, or you can spend a lot of money in the hopes of making more money on a large volume. I should say "there *were* two philosophies," because the first is going the way of the brontosaurus, and I will tell you why.

The government farm programs. We will get more into these later on, but they work like this: the programs are based on acres times yields times monies, and when you are multiplying, you increase the total many times more when you increase one of the factors. Back when a farmer could sell on the open market, he could get by making a small yield if he kept his costs down. But suddenly, after the government had broken the free markets with embargoes, the farmer found himself in a position of having to farm for the government payment. There was no choice, if he wanted to stay in business. And the government has never particularly minded how much is spent on something. I saw a lady on TV the other night holding up a rubber chair-leg tip and saying she could buy them all day long at the five and dime store for thirty-five cents, while the government was paying $1,114 each.

So, the guy whose average was about a bale and a quarter and who had been living quite well off of that, had less to supplement the artifically low market price with than his neighbor, who had been barely getting by on two bales to the acre. Some type of reorientation was indicated, but guess what? The system ain't set up that way.

The ASCS* office, which handles the subsidy payments, says, "Sorry, no adjustments for farming

Agricultural Stabilization and Conservation Services

cheaply!" And the acreage is now locked in. So the only variable left to fiddle with is yield. The cheap farmer goes down to the bank to borrow twice the usual amount so he can get his yields up. But the banker says, "Your history is only a bale and a quarter. I can't cash flow this larger amount!" Now the yield is locked in. And the handwriting is on the wall.

Another philosophy that has gone by the boards is the "we gotta live here" theory. The number one and number two tools of some farmers became, for a while, the bulldozer and backhoe. Ditches were cleared up and filled, then graded back or leveled so that they could be cultivated across. Fencelines and tree lines came down after cattle were sold, and this became cropland. Thickets were leveled; larger tracts of woods were shoved up and burnt. My father watched one neighbor destroy the best dove roost in seven states, and turned to me grimly. "Son, never forget, you gotta live here."

But again, the government programs proved out the neighbor. Total Crop Acreage became one of the things payments were based on. The more cleared land, the more government money. The neighbor might have to go somewhere else to shoot doves and rabbits, but his decision had been a key to remaining in business. People blame the farmer for clearing up the marginal land, and in one sense they are correct. But for survival, it was actually the thing that he needed to do.

A third philosophy that has backfired is the "don't put all your eggs in one basket" theory. A farmer who was used to getting on a tractor and working himself usually diversified in order to spread out the work load. With a thousand acres to work, he might plant, say three hundred acres of cotton, the main cash crop. Then he would have two hundred acres of wheat, two hundred acres of grain sorghum, and a hundred acres of rice. The planting schedule would be milo, then rice, then cotton, then beans. Wheat, planted in the fall, could be harvested after the beans were planted, which was also a great time to have extra money come in. Then rice, cotton, and beans were harvested, and finally wheat was planted.

The problem in this case was that after the government broke the grain markets with embargoes, cotton became the most profitable crop, because of the subsidy payment. But the "bases" for crop acreage were determined by history, so suddenly you found you were locked in to planting only three hundred acres of your most profitable crop. Plant without the program, you say? Aha, we'll look at that in the next chapter.

I don't mean to imply here that only the farmers who farmed close to the vest, left their "waste" land, and diversified, are the only ones in trouble nor that all of them are in trouble. I'm not trying to say that all of the men who farmed mostly cotton, from turnrow to turnrow, and had high yields, are free from worries. My point is merely that some once-valid philosophies of farming have, through factors beyond the farmers' control, made it much easier to lose the farm.

And we're working on plans for the rest of you.

LESSON 3
The Guv'mint

The first sonic boom I can ever remember came about nine o'clock one night in the late fall. An old widower, who lived close to us and worked as our yardman and gardener, had thrown a couple more lumps of coal on the fire and gone to bed. A rat that lived in his attic apparently had a set route of getting to the kitchen. He would squeeze through the hole in the ceiling that the light cord dangled from and climb down the cord and onto an extension cord plugged into the light bulb socket. The extension cord went to a radio on the mantel, and from thence the rat would jump to the door facing, run to the top of the kitchen door and jump to the top of the cupboard. This way he stayed out of the reach of the cat.

The rat was halfway down the light cord when the sonic boom shook the house. He lost his footing and fell to the bed just as Coney sat up at the explosion. The rat lit on the man's shoulder and Coney swiped frantically at the rodent, knocking him into the fire.

The rat hit the coals, shrieked, bounced back to the hearth, and lit a shuck for the kitchen, trailing sparks and cinders. Coney rolled out of bed in the dark and stepped right on the middle of the cat, which could smell singed rat and assumed that was what was attacking her. She lit into Coney's foot with teeth and claws and was rewarded with a tremendous kick that propelled her all the way across the room and through the top sash of the window. Coney, hearing the explosion, rat shrieks, cat screams, and breaking glass, wrongly assumed something was coming *in* the window rather than going out, and headed full speed for the kitchen and back door. The extension cord clotheslined him with such force that the radio was jerked from the mantel and delivered him a solid blow on the shoulder, enforcing his impression of being under attack by unknown beings. The screen door hook held, but the hinges did not, and Coney escaped into the night and appeared on our back steps calling frantically for my father.

With flashlights and shotgun, the two men returned to view the damage, but found no alien beings. My father attempted to explain about jet planes and sonic booms. Coney listened for a few minutes, then interrupted with, "Who dat plane b'long to?"

"United States Air Force," my father answered.

The old man nodded gloomily, understanding now. "Uh-huh. Th' Guv'mint done shuck up de elly-mints!"

Truth is where you find it.

It is my firm belief that most of the financial trouble

connected with farming today is the result of the government shaking up the elements. We're going to touch on several of these elements briefly—I don't intend to be exhaustive or argumentative. I will just simply lay out the facts, leaving out any possible good intentions, and if you arrive at a different opinion, don't bother me with it.

Write your own damn book.

We've already touched on inheritance taxes and how they've forced many farmers into unnecessary mortgages. Let's just cover as painlessly as possible the farm programs, embargoes, FmHA,* and maybe a couple of others.

One of the things that really browns me off in this life is the reference by news media to "farmers feeding at the government trough." At the risk of sounding overly simplistic, let me say that there are only two reasons for this type statement: the person is either stupid or malicious. Well, let's say there is a third possibility: the guy might be stupid *and* malicious!

Most farmers today feel that if the government just opened up the purse strings and *gave* each farmer five hundred dollars an acre, it would not make up for the damage that's been done. These ivy-league jokers still wet behind the ears who are trumpeting that farmers "need to become market oriented" need to do their own homework. If they'd just go back a few years, they would find out that we *were* market oriented.

*Farmers Home Administration

For instance, let's take soybeans. For years most soybean farmers contributed, voluntarily, two or three cents a bushel to the American Soybean Association for research and market development. By the late sixties, we had a lot of the Orient, especially Japan, dependent upon American soybeans and soybean products. Then there was a short crop in 1973 and prices went sky-high, to around eleven dollars a bushel. Consumer groups began to bellyache, notwithstanding the law of supply and demand, and encouraged the administration to embargo soybeans. None could be shipped out of the country. The commodity market closed for several days, and when it opened back up, the price was down around five dollars a bushel. It never recovered. Wonder if the same editors and ivy leaguers were screaming for "market orientation" then?

The tragic result of this action by, I am sure, a well-meaning administration, was that our Oriental customers, seeing that we were an unreliable source of a product they had become dependent on, went elsewhere. They financed or encouraged the clearing of over seven million acres of pampas grass and rain forest in South America that first year and had them planted in soybeans. Now, with over thirty million acres there, they don't need American soybeans to supply the markets that American farmers had so carefully nurtured. Our own government has denied us these markets.

It worked the same way with cotton six or seven years later. Cotton farmers had formed their own research and marketing organizations and had gone into

Eastern Europe and Russia in a big way, especially with blue jeans. Then the Russians invaded Afghanistan. Did our intrepid leaders punish the Reds by denying them toilet paper, or ladies' underwear? Nope. They cut off their cotton. I think our strategy was this: force the dirty Commies to wear polyester underwear and they'll all get jock itch in the hot Afghan climate. A soldier busy scratching his crotch can't shoot, so the whole Russian army would soon be wiped out by Afghan rebels with crossbows and muzzleloaders.

The Cubans crossed us up. Cotton fell from around seventy cents a pound, which the Reds were going to buy at, to around fifty cents a pound. The Cubans, and other market-oriented countries, bought cotton from us for the fifty-cent price, turned around and resold it to Russia at about fifty-five cents a pound. The Cubans made a nickel a pound, the Russians made fifteen cents a pound, and the American farmer lost twenty cents a pound.

To make matters worse, since the demand we had created was still there, other producers stepped into the gap. India, Pakistan, and China increased their cotton production to take care of the Russians and their satellites. How did our government respond to that? Well, they took pity on these poor ignorant third-world countries and sent scientists over there on U.S. tax money to show them how to grow cotton. Now China has gone from importing around five million bales of U.S. cotton to exporting about two million bales of Chinese cotton. And guess who built up the markets they now export to? The American farmer.

So, now we have farm programs to encourage us to become more "market-oriented." The markets are gone, friend; and most of them we'll never get back. Now maybe you are seeing the necessity of the government paying folks not to plant. They ruined the markets we were planting for. But where does the blame fall? On the farmer at the government trough.

Well, let's look at that trough and see if we can understand the slop a little better. The average guy on the street, and any reporter, would say the government pays farmers not to plant. Let's put it another way: the government tells us how much of which crops we can plant, and then compensates us to some extent for usurping that power in this free country of ours.

"But, the programs are voluntary!" you say.

Sure. And so are income taxes.

I'll admit, some programs on some crops in some years are truly voluntary. You can be in them or not, it's your choice. But let's look at how that works. Suppose, on your thousand-acre farm, you have a three-hundred-acre cotton base, two-hundred-acre wheat base, two-hundred-acre milo base, and a one-hundred-acre rice base. One year wheat is worth next to nothing, so you don't plant any. So what happens to your base? Right, it is reduced, unless maybe they let you fallow that acreage. Thing is, suppose wheat goes up to four dollars a bushel next year and you want to plant more than the base. That may be the year the government decides you *can't* plant more than your base. Since most markets are like yo-yos and a farmer can't make long-term marketing plans, you are put in the position of trying to protect your bases in case the price of another crop goes up next year.

Long-term planning is completely missing from government programs. Remember the PIK* program about five years ago? The ASCS offices had meeting after meeting to try to explain it to the dumb farmers. One of the questions that was asked at every meeting I attended was, "If we cut our planting to fifty percent, will it reduce our bases in the future?"

And the answer was always "No."

Guess what? This past year the same people figured *in* the actual plantings for the PIK year in the five-year averages to determine our bases. My base was reduced by nearly fifteen percent.

"Well," you say, "the programs are still voluntary."

Realistically they aren't. If the government has screwed up the markets so that the only way to profit on a cotton crop is to get the payment, then you must be in the program to get a profit. And some years they have cross-compliance clauses, so that if you plant over your allotment in one crop, you lose the payments in the other crops.

*Payment-in-Kind: for instance, a farmer would be given his subsidy payment in wheat instead of dollars for the wheat program.

And then there's the Big Brother theory.

I can't prove this, but I firmly believe it works this way. (If you don't, I repeat: don't tell me. My mind is made up and facts won't change it.) Ultimately, the money loaned to farmers for crop production is controlled by the government, either directly or suggestively. I think that 'way up there in their ivory towers, the bureaucrats say "there is this farm program and we want everybody signed up." The word trickles down the system, and then when you go to see your hometown banker, he says, "You're in the program, right?" Very few farmers in this day and time are stout enough to stand up to their bankers. So if they want a furnish, they're in the program.

There's another reason the banker wants you in the program. Come hell or high water, he's got an assignment on that check. That much he *knows* he will get back. "There are other bankers," you say? Bankers these days go out the back door when farmers come in the front—at the bank, that is. The reverse is true at the farmer's house or shop.

When I'm saying "banker" here, I am speaking of commercial bank and PCA* folks who loan money for crop production. The government is also in the business via the FmHA. At one time FmHA was supposed to be the lender of last resort. This got to be a game for a while ten or fifteen years ago. A farmer who was in good shape but wanted a lower interest loan could go

*Production Credit Association

see a banker golfing or fishing buddy and get the letter he needed to refer him to FmHA. Then hard times began to take their toll and at the same time more farmers really needed FmHA the government began to mandate cutbacks in their loans. The guaranteed loan program was to fill the gap. The commercial banks would lend the money but FmHA would guarantee ninety percent of the loan. Sounds good, right? This would prove to be the finishing straw that broke the camel's back for many farmers.

The one thing consistent about government from the Pentagon to the White House is paperwork. Suddenly, it took farm loans months to be finally approved. Used to be, a farmer could spend a couple of hours in a banker's office, fill out a few forms, and get a commitment, whether yea or nay, in a few days. Now that government was involved, it took a couple of weeks just to get all the forms filled out. And farmers, as a species, are not particularly charged up about paperwork. A clerk in St. Louis always sent some form back to be redone. And everything had to be approved on infinite levels. Many times a farmer had all his crops up and growing before his production loan check ever came in. Being out on a fifty-or sixty-thousand-dollar limb really increases the pucker factor, believe me.

So many farmers started delaying their planting dates, waiting for a check to come first. This was fatal in years when we had early fall rains. Quick approval—or disapproval—should have been mandatory. Farming decisions had passed to people whose closest association with agriculture was the onion or olive in their martini. They apparently could not have cared less whether you plant cotton in April or June. But you'd better have all those forms filled out in triplicate!

I'm sure we both could name a lot more ways that the "guv'mint has shuck up de elements," but it seems to me it's time to put the chairs in the wagon as far as this lesson is concerned. Answers? Well, I've told you before that I'm only dealing in problems here. I do have a suggestion, though.

Let's turn it around a few years. Since the farmers are going broke because of the bureaucrats, let's put the farmers in charge of the government farm-related agencies and put the bureaucrats to farming the land that the government and banks are foreclosing on. Would that be too cruel?

You heard about the Delta farmer who was jailed for child abuse last week, didn't you?

Yeah, he gave his son a hundred acres of farm land!

Lesson 4
The Lenders

"A penny saved is worthless unless it is spent." Matt Prewitt, Greenville, MS, 1986.

Unless you get the wrong impression, let me hasten to say that I don't necessarily condemn all bankers just out of hand. Heck, I graduated with a degree in banking and have wished many times that I had gone into the business. One of my best friends is a banker—the famous Dude McElwee. But, in retrospect, I want to discuss how some of the lending policies have now jeopardized many farmers and ruined others.

The late sixties and early seventies, before the grain embargoes, were, in general, good years for most farmers and their banks. This was a period when many young farmers were getting into the business, for several reasons: land was still relatively cheap; at the time there was good money to be made farming; the government, with minimum wage laws, had practically outlawed the sharecropper system which had provided cheap, plentiful labor for so many years, so farmers were turning to bigger equipment, which encouraged expansion; and finally, the "baby boom" kids were coming home from school and service. A kid who had been raised on the family farm and was buying into a going concern was considered a good credit risk. As I said in an earlier chapter, there are few other businesses where a young man can get into so much debt so quick.

True, no one held a gun on these farmers and made them borrow the capital required for this period of expansion. But no one held a gun on the banker, either. Twice, I had Federal Land Bank men ask, "Are you sure that's enough?" The rule of thumb seemed to be "go ahead and get what you want," and many of us did. In figuring the payback, though, nobody foresaw what government interference would do to us. Everyone will admit, at this point, that there were some farmers who abused the system. How they got by with it so long always amazed me. I know guys who would get their furnish loan, buy a new truck and car, buy a new ski boat, and take a trip to Las Vegas—without their wives. (To me, that was the greatest mystery of all.)

Trouble was, this became a popular image with non-farmers. Gone was the perception of the somber sun-wrinkled man in overalls with a pitchfork in his hand; instead the general public thought of farmers as irascible playboy types who were not ashamed to flaunt their wealth. A few bad apples were beginning to taint the barrel.

(I believe the same thing happened nearly a hundred and fifty years ago with Southern farmers. The public was led to believe they all had slaves and whipped them all soundly before they sent them to bed. Nobody paid much attention to the fact that most farmers didn't own slaves and the ones that did usually didn't beat them. Today's parallel to that: how many farmers do you know who buy a fifty-thousand-dollar tractor and then work on it with a sledge hammer when it breaks down? Besides me, I mean.)

Still, the image persisted. And, in the latter seventies and early eighties, we began to pay for it. Farmers who had begun to have a tough time would sit down with their banker and make out a budget just as bare bones as they could sincerely get it. The lender many times would then cut the requested production money by at least ten percent simply because of the image. This meant that an honest farmer had to cut somewhere, so now what happens?

He skimps on land preparation, drops off on some of his subsoiling, reduces fertilizer and chemical rates a little, stretches his seeding rate below recommended levels, allows insect levels to get a little higher before poisoning, skips a poisoning or two, tolerates more Johnson grass or cocklebursin his fields, gets by a little longer on equipment that needs replacing. In short, a good farmer is forced into becoming a sorry farmer.

The natural result of such shoddy practices is a loss in yields. When the farmer goes in for his furnish loan the next year, his lender will cast a stern eye on the yield figures and cut the production loan again. Now, the handwriting is on the wall.

A sequel to this attitude was the time involved for

approval of production loans. We have already discussed this in the previous chapter, so let's just say here that no lender did a farmer a favor by considering for weeks or months over his production loan and then finally approving it after the guy should have already had his crops planted. In general, Mother Nature's rule is: you plant late, you harvest late. A farmer who is put in the position of having crops still in the field when the fall rains start is up a long creek with a short paddle.

One wet fall one of my neighbors and I parked next to each other on the gravel road and commiserated about our muddy, rutted fields. Gloomily, he said, "I'll bet it's even muddy in hell right now!" Right in behind that statement, he declared, "I'm sick of rain and mud! I'm leaving!" As he cranked up and drove off, I assumed he had headed home. A week or so later his landlord drove out and asked, "You seen Bubba?" I heard they found him three months later on a beach in South Florida. He was broke, but dry.

I wished he had asked me to go with him.

One of the lenders most farmers feel the hardest toward in this area is the Federal Land Bank. Most nonfarmers feel that the name of this institution means that it is government-owned, but that is not the case. It was started as a farmer-owned/stockholder-type organization and quickly paid back its government seed money decades ago. It is now in danger, however, of either going under itself, or being voluntarily taken over by the government. Well, as another farm boy author, Ferrol Sams, said in *Run With the Horsemen*, "If she's ruint, she done it herself!"

For if the Federal Land Bank had shown the same consideration for its own stockholders as it has shown for total strangers, I feel that many busted or near-busted farmers could still make a go of it.

Consider: most of my own land mortgages are with the Land Bank, therefore I own thousands of dollars (like maybe twenty thousand) of stock that I was required to buy in order for them to approve my loans. My interest for the last several years has probably averaged around thirteen percent. I made the decision several years ago to sell some land to relieve some of that interest and debt load. Two potential buyers backed off of buying my land, which sale would have put me in better position as an FLB borrower, because they could buy FLB foreclosed land cheaper and at a lower interest rate.

A check with Land Bank officials confirmed that this was true. While a stockholder in fairly good standing for twenty years had to pay thirteen percent interest, the FLB would sell foreclosed land for five to ten percent down, no stock required, eight percent interest for the first five years, and no personal liability for the first five years. No, they would not finance a man to buy *me* out on those terms. No, they would not take *my* stock in as a credit on *my* yearly payment and give *me* eight percent terms.

With these "sweetheart" deals, they rendered their own stockholders' lands almost unsellable, drove land values down, and ensured that many of their own long-time customers and stockholders would go down the drain.

"Why would they do that?" you say.

I honestly can't figure it myself. But I have heard smarter people than I am say it was because they desperately needed a bail-out from the government and felt they should appear to have moved some of the foreclosed land in their inventory. I might ask, though: at just what point do you consider throwing the baby out with the bathwater?

One last point on lenders. In these past several years, when folks who had been farming for decades began getting in financial trouble, we farmers started comparing notes and found that many farm lenders, some of whom were not five years out of college, had figured things out. They began to advise farmers on chemicals, fertilizers, seed, irrigation, pricing, and other details of farm life the farmer himself might have overlooked. (Many farmers, on the other hand, were upset at this presumptuousness and freely gave the offending lender advice on his sex life.)

During this period, my banker called one late spring day and announced that cotton futures had gone up to sixty-two cents. He "strongly suggested" that I run right down and sell at least half of that fall's crop. Frankly, I was in no position to turn down a "strong suggestion," so I tricky-trotted right down and sold half my crop for sixty-two cents.

That fall I sold the remaining half for seventy-eight

cents. When I reminded the guy, though, he didn't offer to make up the difference. "You know what they say in the Russian Marines," was his attitude. He did not advise me on prices again, though.

I told this story to a lawyer once. He was aghast and sputtered, "Man, haven't you ever heard of lender's liability?"

Verbal, man. All verbal.

Fellow farm losers: if you don't pick up on anything else in this little book, get one thing down good right now.

For God's sake, get it in writing!

Lesson 5

The Good Years

"Oh, Boy, finally something good," you must be thinking. Yes, there really were some good times. Most of the sixties and the early seventies brought not only prosperity, but fun, farming. The government had cut out Uncle Remus, but labor was still plentiful and strains of "zip-pe-de-do-dah" could still be heard around the plantations.

I broke in about the middle of this period, and, as best I can remember, we had around a dozen tractor drivers, a farm mechanic and manager, and myself, my father and my uncle to actually plan what work needed to be done, who would do it, and how it would be done. In general terms, my job was to go to town for chemicals, seed, parts, and so forth; calibrate the spray rigs; tell the manager if something needed fixing; and knock folks off at dinnertime and dark.

I was wistfully remembering this when I met another young farmer at the elevator about nine one night. I had been on either a tractor or the combine since six-thirty that morning, with half an hour for lunch, and an hour to get one of the other three drivers unstuck. I had knocked off the others and taken the last truck of beans in myself. The elevator was a block from the fried chicken place, and the aroma was mouth-watering.

"You know," I said mournfully, "I can't remember my Daddy ever even getting on a tractor!"

"Hell," the other man snorted, "I can't ever remember mine with his britches dirty!"

It used to be that way for lots of farmers in the Mississippi Delta if you can remember back that far. The period of prosperity led to a period of expansion. As we have noted, the old sharecropper system had been forced out of existence by a well-meaning, but ignorant, bureaucracy. Instead of the old smaller tractors and four-row machinery, the farmer went to larger tractors and six-row equipment. With larger machinery to justify, folks looked around for places to expand, and bought out Widow Jones's place or Old Joe's place or maybe just logged and cleared up some woods and low sloughs the Corps of Engineers had drained.

Land was not readily available, but what was available was still relatively cheap. Same with rental land. Some of the older farmers who didn't have sons or son-in-laws coming back to the land retired after their sharecroppers had to, rather than invest in the bigger equipment now called for. It was not uncommon to hear of ten-year leases for, say, thirty dollars an acre, being locked in back in the late sixties.

It is my understanding that we were behind the rest of the nation by several years at this point. In areas around the country, especially farmland affected by urban sprawl, land values had already dramatically increased, and one of the factors involved in driving our values up was a migration. People in Missouri, for instance, could sell their old granddaddy's five-hundred-acre farm for three thousand dollars an acre, take that money and move south, and buy three thousand acres for five hundred dollars an acre.

Another commodity in good supply at this time was willing, responsible, dependable labor. Though the sharecroppers were no longer able to raise their own crops, many of them still lived on the farms and had driven tractors for years. They were loyal, knowledgeable about what needed doing and when, and they would take care of their equipment. There were a few "hot-rods," but their older peers usually settled them down. The next ten to fifteen years would see most of these old gentlemen pass on, but fond memories will always linger of those subservients in title only, who would roll their eyes and sigh remorsefully when having to correct the "Little Boss."

Prices were relatively good during this period, compared, of course, to the prices you had to pay for farm supplies and equipment. Nobody had to borrow a lot for their production money and that was readily available when you needed it. Many farmers followed a routine of this nature: they would pay off their production loan, or "furnish," out of their cotton crop, then the tail end of the cotton and all the soybeans would be gravy; any capital payments would be made from this surplus, bonuses would be taken out, and the farmer would still have enough left over to start a new year on; along about April, the farmer would make arrangements for his line of credit for production purposes and would use this borrowed money until the wheat was harvested in late May or June; production money would be needed again in late summer, but then early rice, milo, or beans were ready for harvest, after which the cotton picking started and furnish money was soon paid off and the cycle started again.

Farming was undergoing a major industrial revolution during these years. More powerful and bigger machinery cost more but meant that the job would be done quicker. The cost of this machinery, however, clearly dictated that it be used enough to justify the price. As we've noted, expansion was the answer. The money was available, the farming was good, farm technology was growing by leaps and bounds, our markets looked stable. What could go wrong? At that time, it looked like farmers had the world by the tail. What we thought then was the world, though, turned out in another decade to be a tiger. And in addition to not being able to let go, many of us have found that the tiger has taken a large chunk out of that rearward portion of our own anatomies.

Bad thing is, he still looks hungry!

Lesson 6

Breakin' Even

One of the problems farmers have had down through the years has been their independence. In a world that was becoming more and more government-controlled, and that government influenced to a great extent by lobbying efforts, farmers should have looked to banding together for the best interests of the industry as a whole. To be sure, efforts at marketing associations worked well; but there were too many different kinds of farming, and nobody successfully combined forces of ranchers, dairymen, cotton, rice, soybean, and rutabaga growers. At a time when farm populations were falling from maybe 20% to less than three percent this was to prove costly.

I have never observed this experiment, but I believe the estimate. If you were to lock ten farmers in a room with one thermostat, you could expect the following results: two would be too hot; two would be too cold; two would be just right; two would begin to argue about the weather outside the room; one would just want out; and one wouldn't give a damn. And since they're all used to being their own bosses, mayhem would result.

I believe I could get a consensus on this opinion, though: at least in our area, the end of the really good times was in 1974.

Several reasons combined to make this so. First, of course, was the 1973 embargo. That, in turn, broke the good prices. This year also marked a run of several years during which crops were, with few exceptions, average or below average. Interest rates were skyrocketing, as were prices for the things we had to buy. The cost of farming was going up, but farm income couldn't keep up.

Weather-related conditions also plagued some farmers, and a period of belt-tightening was indicated. To a degree, this was not possible, though. You remember the old song about the old lady that swallowed the fly, don't you? As I recall, she then swallowed a mouse to catch the fly, swallowed a rat to catch the mouse, swallowed a cat to catch the rat, swallowed a dog to catch the cat, swallowed a horse. . . . you get the picture. The last line of the refrain says "I guess she'll die."

In general, farming will too, as we remember it.

For in these Breakin' Even years, when we needed to stop, we found it wasn't that simple. There was always something left to swallow. The expansion of the good years had to be paid for.

Those who had bought in during the last of the

sixties and the early seventies were introduced to the variable rate interest loan. We found that it was called this because the interest rate was to indeed vary considerably during the next several years, up to nearly twenty percent. This meant that a man who might have bought, say, five hundred acres at four hundred dollars an acre financed saw his yearly interest payment shoot up from $120,000 to nearly $400,000. Now expansion didn't look like such a hot idea.

Folks started cutting back where they could. Since the bigger machinery required less labor, and labor prices, mandated by a government that failed to take into account the many fringe benefits of plantation living, were going up, farmers began letting not only laborers, but managers, go. This equipment had warranty on it, didn't it? The implement dealer's service team could do the job for less than the cost of keeping a full-time manager-mechanic. And, historically, bad times price-wise, weather-wise, or otherwise, were temporary. By the time warranties ran out, the pendulum would have swung back, and new equipment could be obtained with warranty in place. Heck, all a farmer had to do was know how to change the oil or tighten a belt.

Until the pendulum didn't swing. Now we are teaching ourselves and our drivers the basics of fixing machinery. Why, just a year ago, I had spent the morning (1) adjusting the brakes on the picker; (2) changing a solonoid on a tractor; (3) repairing a bent bar in the picker; and was now (4) replacing several broken teeth on the combine blade before driving it to the field myself. I had just mashed my thumb on the next-to-the-last tooth when the picker driver walked out of the field and up to the barn.

"Mista' Bob, de rear wheel got plumb cross-ways in de mud an' I couldn't tell it, an' de lugs pulled slap' out. Hit's settin' down in de middle of de field!"

It only took me one exasperated glance to determine that Herbert's assessment of the situation was correct. A quarter-mile down in the muddy field the picker sat back on its haunches at an angle, seeming to thumb its nose at me. I took a deep breath and gave precise instructions.

"I am going to get on this combine and go the bean-field. Get Tommy off the bean truck and take a tractor, hook up to the little trailer, put everything you need out of the shop on it, and get ten gallons of diesel. If you fix it, fine. If you can't, pour diesel on it and burn the s.o.b. If I see smoke over the treeline, I'll know you couldn't fix it. But whatever you do, don't bother me with that damn picker again today!"

Worked just like a service manual. They had it running in three hours.

It was during this period that farmers found that their previous expansion in farm areas had caught up with them in the service areas. It was a truly marvelous thing to go from a four-row combine to one of the new six-row machines. Not only could you cut half again as many beans in one swath, but the newer machine operated at a faster speed with less choke-ups. Now, however, there was another problem: what to do with the beans. My old truck that had kept up with the old combine not only couldn't keep up with the new machine, but since most other farmers had newer combines, too, the lines at the elevators were longer.

The solution was evident. Either I had to increase my trucking capacity, or build on-farm storage bins. Either solution meant more capital investment.

Same thing happened in cotton. Bigger, faster pickers meant going to more and bigger trailers, and then module builders; both of which entailed more investment at the cotton gin. Not that many years ago, this area was served by over a dozen cotton gins, and now there are only three who accept customer ginning. They are all big high capacity co-op gins owned by farmers who put up the capital themselves.

Elevators have followed the same trend. So have catfish processing and food plants, and other farm-support industries. This has all required investment capital, though some of these investments have paid back well over the years.

All this put an extra strain on farmers during a period when, in general, they weren't making big profits anymore because of prices, weather, interest rates, higher costs, and capital outlays. But the attitude was "we'll be okay when the pendulum swings back."

Farmers have always had to endure lean years. At the

coffee shop an older retired farmer said after a second so-so year, "Don't worry. I've been farming since the '20s and I've never seen three poor years in a row."

After the next year, he said, "Don't worry. I've been in farming since the '20s and I've never seen four poor years in a row."

After that year, he quit coming to the coffee shop.

Lesson 7
The 'Flations

I had a dream the other night. In this dream, I took a truckload of soybeans down to the elevator, weighed it, and told the man, "I'll have to get $9.50 a bushel for those beans. If you want them, cut me a check." The guy just fell all over himself getting to his checkbook. I then took the money, went down to the car dealer's, and said, "I'll take that red convertible. I'll give you a thousand dollars for it, less two percent for dust on the top. Take it or leave it!" The guy just fell all over himself making out the bill of sale.

Then I woke up and found that in real life, things are just the opposite.

As we have found from our previous discussions, government embargoes have been the single most damaging factor in deflating farm product prices. Add to that the example of our tax money going halfway around the world to teach someone how to better grow agricultural products to better compete with American farmers in the marketplace that Americans built up, and you have grasped the problem of commodity prices. The American farmer has to take what he can get regardless of whether he can make a profit on it.

In the past twenty-five years, cotton prices have gone from thirty cents a pound to ninety cents and are currently back around thirty cents. Soybean prices have gone from around three dollars a bushel up to eleven dollars and back down to about $4.50. Wheat has been as high as $4.50 a bushel during that period, and now runs around $2.25. Rice, corn, milo, cattle; same story.

But everything is relative, we all know that. A man can afford to take less for his product if at the same time it costs him less to produce it. Even an idiot can understand that it matters not if you get paid twice as much to work in Alaska when it costs twenty dollars a dozen for eggs and everything else follows the same pattern. So since the story on farm products is deflation, what is the story on farm costs?

Inflation. You knew that, didn't you?

Land taxes during the past twenty years on my farm went from around $3,000 to $11,000. A regular size pickup truck in the past ten years has tripled in price. A medium-sized row crop tractor during the same period has gone from $10,000 to nearly $40,000. Granted, that new tractor will do more work than the old one, but not four times as much. One of the better soybean herbicides went from about seven dollars a pound to over sixteen dollars.

Labor? Well, right after the sharecropper system went out the window, it seems to me that my father was paying tractor drivers twenty-five dollars a week, and that was higher than the usual farmer. Last year I was guaranteeing my best man $133 a week. That's "straight time," of course. On an hourly basis, without the fringe benefits involved, and with the overtime, figure labor has tripled.

During the past twenty-five years in this area, farmland prices have gone from three hundred dollars an acre up to over fifteen hundred dollars and then went the other way back down to who knows what now? Interest rates on that land during the same time have gone from six percent to nearly twenty percent and hover around eleven percent right now. Production loan interest is the same way.

Speaking of production loans, look at present USDA estimates on the cost of production: three hundred dollars an acre for cotton, three hundred dollars for rice, seventy-five dollars for beans, wheat, and milo. Yet I can remember borrowing a little over fifty thousand dollars to farm two thousand acres. Now production loan interest alone would run me half of that. And land payment interest is nearly triple that.

The point of all this is not just a dry recitation of figures. The point is that, in an age when nearly everything else is going up, farm prices are going down, and the farmer has no control over that. When the price of newsprint goes up, the newspaper raises its price.

When the price of steel goes up, the car maker raises his price. When the price of owning a house goes up, the landlord raises the rent.

I was on the road last week selling newspaper columns. An editor liked my stuff, and said, "How much do you want for this?"

I went into shock. It was a completely alien idea to me, to be asked to set a price for my product. I stuttered, stammered, blushed at my own damnfoolishness, and finally said, "Would fifty dollars a quarter be too much?"

"That's too cheap," he said. But he cut the check for fifty anyway.

Ironically, other agricultural areas have financial problems. The economics of two of our biggest soybean competitors, Brazil and Argentina, have been on the brink of failure for years. If they did fail, American farmers would be much better off in the international marketplace. But who keeps these foreign competitors going? The American government, through the World Bank and the Export-Import Bank.

To quote Pogo, "We is met the enemy, and they is us!"

See if you think a non-farmer can understand this statement: our own government has forbidden us to sell our products overseas on the free market, thereby encouraging other countries with cheaper costs to enter that marketplace and supply the demand that we built up; our government has then taught these foreigners how to increase their productivity of the same commodities we grow here; when the economies of these countries threaten to fail, our government keeps them afloat and thereby in competition with us; and then our government tells us that our problem is that we overproduce!

"Farm bill costs too much!" The politicians and media trumpet on page one. Yet on what page do you read how much money those same politicians spend to keep competing nations' economies alive? And on what page have you *ever* read that the government sends scientists to help foreigners compete with American farmers?

Can you imagine this scenario? On Monday night football, the Bears sack the Broncos' first-string quarterback on the first series of downs and he leaves with a broken wrist. The second-string signal caller has the flu. The third stringer lasts until the middle of the second quarter, when a blitzing linebacker rings his bell. Then the Bear coach runs out on the field and calls time for a uniform change. "Here, take one of my quarterbacks," he offers to the Bronco coach!

Right. This only happens in farming.

Speaking of uniforms, the 'flations have struck in the final consumer end of farm commodities, but the farmer didn't get the benefit. A cotton shirt that costs twenty-five dollars has about a quarter's worth of cotton in it; a dollar loaf of bread has about three cents worth of wheat in it; a pound box of farm-raised catfish selling for three dollars the farmer got maybe forty cents for. And so on ad infinitum.

Once a year, I try to give a special present to my wife:

I go shopping with her. We spend all day in and out of stores looking at stuff and I never get mad, tap my foot, yawn, or look at my watch. A while ago during one of these excursions, we wandered into a shop that sold only lingerie. I felt like a cat in a shopful of canaries, and began to pick out items that needed modeling.

There was one very attractive outfit there of cotton silk, that only covered part of this and very little of that and hardly any atall of the other. Every home ought to have one, I decided, and since it's all cotton, I knew I could justify and afford it.

Fellow farmers: if parity is ever brought back, I now know what product to base raw cotton prices on.

And what man would argue?

Lesson 8 The Collateral Crunch

(To be sung briskly)
The papers on equipment
You sign right after crops,
Then they want the land
Which is where you ought to stop.
Then they want the house
Then they eat your lunch.
And now you are dancing
The Collateral Crunch!

Not quite the peppermint twist, maybe, but there are an awful lot of folks dancing to it.

Because of the interaction of the 'flations that we studied last chapter, a great many farmers are threatened by a phenomenon called The Collateral Crunch. "Crunch" as in between two hard objects. Those who are not crunched are getting squeezed, to say the least.

It seems like a long, long time ago in a place far, far

away, that a farmer could secure his production financing by using his crops and his equipment for collateral. Several years ago this began to change, at least for a lot of farmers who owned their land. It is now commonplace for farmers to have to put up their land as security for their production loan, or furnish. There are several reasons for this, and I understand most of them. One reason, though, is entirely my own speculation, and if it is true, I don't see any fairness about this policy that is readily apparent.

Bear in mind that this is entirely speculative on my part, and I don't intend to try to document it. Just put it down to a feeling. Kinda like when you are talking to a good-looking girl at a party and suddenly you just know your wife has walked into the room—even though your back is to the door.

I firmly believe that a great many of us who own our land have been the victims of a very subtle averaging policy by some farm lenders over these past few years. I think it worked like this:

Bill, Fred, Joe, and Al all farm about a thousand acres each out east of town. Bill and Fred own their land, while Joe and Al rent theirs. The area has a couple of years of bad weather, and the bankers begin to get itchy. So far, the four men have only pledged their crops and equipment as collateral for their production loans. Now, however, the lenders want more security, but Joe and Al have no more farm-related collateral.

Aha. Bill and Fred do.

On the surface of it, Bill and Fred are required to put up some of their land as additional collateral for their own operations that next year. Joe and Al, though no better farmers, continue to get their furnish for crops and equipment pledged. I believe the philosophy was to hedge the bets; if they all go down, at least the four thousand-acre operations are secured by two thousand acres of land. And that's the way it's working, beltwide. You don't believe a word of it, you say?

Fine. Just don't waste *my* time with opinions like that.

Whether you agree with the Collateral Averaging Theory or not will not affect your final grade here. But certain facts are indisputable, and you can be sure they will be on the exam.

Fact: crops and equipment used to be enough collateral for furnish loans.

Fact: crop prices fell, reducing crop values as collateral.

Fact: as times got harder, farmers started stretching their older equipment, reducing its value as collateral.

Fact: as more farmers began going under, more used equipment became available, reducing the value of equipment farmers owned as collateral.

Fact: equipment dealers were becoming desperate for business, too, so we began to see "100% financing" of new equipment. But the equipment dealer held a first mortgage, so the banker's position was not as secure.

Fact: the farmland was the next thing the lenders

could look for to secure their production loans. Only renters didn't have any. Some of them *were* asked at this point to put their houses up, though.

When the Collateral Crunch first began to rear its ugly head, farmland values were not only still high, but they were still rising. A farmer might already owe six hundred dollars to eight hundred dollars an acre, but there was room for more. A bad crop? Roll over what's left on the note, add it and the land payment onto the mortgage, and go right ahead. Sometimes this was not the thing the farmer necessarily wanted to do, however. But again, remember our discussion on "easy"? Once you've taken the step, it ain't easy to stop before you hit bottom.

One year I ventured up to "kneepad corner," as we called the waiting area for farmers at the bank. The guy in front of me was talking to one of my neighbors, who was next in line. This neighbor was very definite about his plans for the coming year.

"I'm through!" he declared emphatically. "I'm sick and tired of lying awake at night gritting my teeth and worrying. I've got a couple hundred acres my daddy left me that's still unencumbered and I'm gonna get a job, plant a big garden and maybe some milo to hunt doves over, and learn how to sleep all night once again."

He was called into the banker's office. Full of resolution, he squared his shoulders and marched in.

A half-hour later, he shuffled out, looking bitter, older and hopeless. I stood to shake his hand. "Was it that hard to quit?" I wanted to know.

"Hell, I couldn't quit!" he said listlessly. "The guy said it don't work that way. he said I still had something they didn't have a lien on. I put up that last two hundred acres." He turned at the stairs. "You reckon they'll let me quit when that's gone?"

Falling land values were the key to the situation we're calling the Collateral Crunch. All of a sudden, in a couple of years' time and for reasons we've already covered, land values began a dramatic fall and haven't quit yet. Lenders who had willingly loaned maybe a thousand dollars an acre on land they valued conservatively at fifteen hundred began to realize that land prices were coming down fast and a bunch of loans were beginning to look not so hot. They also realized someone was going to have to take the blame, and blame is not something that's willingly shared.

In the Navy, I became familiar with a game called "cover your own . . . rear end." I was wrong to conclude that the basics of this game were known only to those employed in government service.

Some lenders learned quickly to play this game, and covered their rears with more collateral: land, houses, stocks, other businesses or investments. I knew a guy who had a current loan whose banker called him in and informed him that his collateral property had decreased by half in a year's time, and unless the man put up his house, he would not renew the note. This was one of the cases I referred to earlier in which the customer gave the banker rather intimate and, to be quite objective about it, unperformable instructions. One word led to another, and sure enough, the banker foreclosed. The man still has his house, though.

I am convinced that many of us put up land as collateral when it was not really advisable or even necessary, because 'most everybody was doing it. This was probably a turning point when we should have gotten good legal advice before doing what we did. Many a farmer is in sorry shape right now because he was used to doing business on a word and a handshake; but these are not valid tools in the game of "cover your . . . rear." Nowadays, a farmer needs to take a lawyer and an accountant with him to get his furnish loan, and he damn sure better read the fine print.

You know what I hate most about this situation? I hate that it symbolizes the end of a way of life in which a man took pride that his word was his bond and that written contracts were for people who didn't trust each other. I saw my father refuse to do business with a broker one time because the man questioned his habit of keeping records in pencil. "Man that's got to have everything written in ink and signed ain't to be trusted," he declared.

But, I've also seen an attitude among farmers themselves that is exemplified by comments like these: "Broke down? Get what you need from my spare parts pile. I'll need something from you down the line;" or "I finished yesterday; you need a few days help before I put my machines up? Just put in diesel and pay the driver;" or "Borrow my rig. I'll get it back when I need it." A group that has been really *too* fiercely independent is now pulling together through necessity.

Remember General Nathan Bedford Forrest's key to success? "Git there fustest with the mostest!" Well, I'm afraid these farm efforts are just the opposite.

Too little, too late!

Lesson 9

Disaster Years

Remember the old joke about the woman who backed into the airplane propeller? Disaster! Suffice it to say that there are lots of farmers who feel the same way about backing into the weather these past few years.

Farmers have always had to contend with the weather, but that's what you plan on up front. There ain't a damn thing you can do about it, so you come prepared to react to it. If it's dry, you go to the field and do what needs doing. If it's wet, you get the equipment ready to jump in and perform the next needed operation. When a prolonged spell of weather sets in, be it drought or rain, you pull in your horns and try to make the best of it. As my father used to say, "Do what you can when you can; and when you can't, go fishing!"

It was a rule of thumb that the weather was going to hit you a real lick once every eight or ten years. By the same token, once every eight or ten years, everything would be just right the whole year and you'd make a ton of money. The other years would in general be more good than bad; and since cotton was to some extent a dry weather crop anyway, the main cash crop was not particularly susceptible to the vicissitudes of the weather.

I'm not sure who to blame for it, unless it's the government, but weather patterns seem to have changed drastically this past decade. If a farmer could plan his weather, it might be scheduled like this: dry February and early March to get the spring breaking done and set up rows; rain the last of March and early April to settle the beds and let a fellow hunt wild turkeys for a few days; dry during the week the rest of April and early May with a slow inch of warm rain each Friday evening to sprout what's planted and activate chemicals; warm the rest of May with same rain schedule—matter of fact, we can extend that Friday evening rain until September 1 (Mr. Ugh McGee used to say, "an inch of rain every Friday afternoon will make a good farmer out of a fool!"); plumb dry during September and October and early November until harvest is finished and fall breaking has been done; a good rain to sprout the wheat, just enough dry to get a stand, then normal winter rains until February.

Now, it seems we can plan on a cold, wet May; an extra dry, hot summer; and half our yearly rainfall during September and October.

Weather-related disasters have marched by on each other's heels during these past eight years. In our area, for instance, there were horrible droughts in 1978,

1980, and 1981. 1979 was excellent, but 1982 and 1983 were marginal. Then 1984 and 1985 hit, with twenty-two inches of rain in one October and seventeen inches the next. The drought of 1986 followed, with no rain from the 10th of June until the last of August and three weeks of 110- to 115-degree temperatures. Six years out of nine brought us disaster weather from Old Mother Nature.

What made it even worse was that, in 1984 and 1985, the crops were really good. In '85, for instance, I didn't ever run the center pivot irrigation equipment during the summer. My rain gauge on that part of the place showed thirty-five inches of rain during May, June, July, and August. That was unusual, of course, for a mile away at my house, I had only twenty-two inches during that time, and another mile west my neighbor had only ten. Prospects looked great until the rains continued through October and November. There are few things more frustrating than harvesting a good crop in the mud. You get a half to two-thirds of what's actually there, and the grade on that is poor. Before the October rains started, I was combining in a 150-acre field, getting thirty-six bushels of beans to the acre with no dockage. When I got back into that field after a month of rain and warm weather, I was cutting nineteen bushels to the acre, but taking sixty-five percent dockage because of the mold and mildew! It was not worth running the combine.

When your troubles can be blamed on a neighbor, your implement dealer, a lender, or a government official, you can at least stick your tongue out at him and cuss him out—even if it's behind his back. Most of us have found that there's no satisfaction in these tactics when the weather is the offender. Matter of fact, some of us are of the opinion that it may make things worse.

Weather disasters are bad enough on production loans, but they make it doubly bad if a farmer is buying his land and has, say, a fifty-dollar-an-acre land payment due the first of the year. When you can't quite pay out your furnish, then there's nothing left over for land payments. The only solution is to double up in one way or another—either borrow extra in next year's furnish loan or rework the mortgage to include this year's deficit plus the due land payment. The land payments on my place more than doubled during this period, because of having to "roll over" yearly land payments into the mortgage.

A sympathetic (or really, empathetic, bless his heart) banker once told me he thought a good solution to the farm debt problem would be to take into account this type disaster on land payments by simply extending the note by a year. In other words, if a farmer has an existing mortgage requiring a fifty-dollar-an-acre payment for the next twenty years, and has a really bad year, just take that fifty dollar payment and tack it on the end, for the twenty-first year. The present system requires a reworking of the mortgage, and you end up with a sixty dollar payment instead of fifty dollars. With a string of years like we've had, many farmers' payments have jumped from fifty dollars an acre a year to nearly one hundred dollars. I don't know if my friend ever got his idea across to the big shots or not, but I admire him for having the guts to suggest something new instead of just playing the old "cover your . . ." game we have discussed before.

I'm going to get serious for a moment. I want to talk about an attitude I have noticed among farmers after these years of weather disasters—an attitude I know is wrong, but one I don't know how to shake. (However, my part of this book is problems; Jim Baugh is in charge of solutions, and maybe he can come up with one for this. The least I can do for a man with his credentials is get the problems defined, though.)

The attitude I mean is a tendency to blame God or, at the very least, to figure He's got it in for you personally and farmers in general. So at the very time a man needs to draw closer to God, he ends up drawing away from Him, almost scared of Him. When a man begins to believe that God Himself is against him, he is very close to the end of his rope.

When a man is in this shape, he badly needs some tangible evidence that God cares for him. Whether that evidence comes through a banker, an implement dealer, a sky pilot, a pharmacist, the couple down the street, or the family Labrador—well, he just needs it, that's all. Man, it ain't so that God's mad at you, but I don't know how to prove that to you.

But I do know this: if you can hang on to your faith and make it to Heaven, you won't have to pick cotton in the mud. Promise!

Lesson 10

Between A Rock and A Hard Place

Where do you start? Well, for the sake of brevity, let's just start at the end. Like the title says, between the rock and the hard place.

I'm going to tell a story here. I'm going to cut out names and try to soft pedal the bad parts, but if you haven't gotten to your own final chapter, I want you to know how complicated and frustrating it can be. And how senseless.

For starters, let me say that, there were four creditors involved. One was a commercial bank, another a member of the farm credit system, and the other two were government agencies. That's a can of worms in itself, and what happens if you get in this situation is that they all usually want more pieces than there are in the pie. Let's just call them lenders one, two, three, and four.

Lender one, after the wet falls of '84 and '85, strongly suggested that I rent part of the land out. Since a partner had bailed out between those years, I heartily concurred, and managed to rent about six hundred acres, including all the cotton and the pivot irrigation system, for a good price—cash in advance, payable in December. I also lucked up on a guy who wanted to buy thirty acres for nine hundred dollars an acre on a piece of ground I owed about six hundred dollars an acre on. He signed a contract in September '85 and put up the earnest money. That made $75,000 I would have available that fall for land payments.

To complete these transactions, I had to get either mortgage releases or crop non-disturbance clauses from my lenders, plus the Internal Revenue Service. "Uh, oh, you named somebody and you're gonna get in troubulll!"

Nope, I did it on purpose. You know who was the first one with a release? Even called to say the letter was on the way and good luck? The IRS! Maybe this isn't the popular view, but they were sure Johnny-on-the-spot for me then. One by one, the other lenders got their releases in over the next six weeks.

Except Lender one. He would give neither a release nor a non-disturbance clause.

This rocked on all winter; they were going to, then they weren't. The would-be renter and buyer both did their land breaking and planted their crops on the assumption they would be able to pay their money. I began making preparations for my own furnish. One lender told me I might be qualified for a low interest disaster loan and sent me a bunch of forms to fill out.

Sure enough, I qualified. In the words of the man, "If you meet this criteria, and you do, this loan is automatic. There are no contingencies. We are just a conduit for this money."

Do you remember the moral of Lesson Four? Get it in writing! I didn't.

Well, we filled out all the little forms and just as I was leaving, he called me back. "Wait a minute," he said, "seems like I remember something else about these loans. Let me look something up." He spent about twenty minutes going through some of his books and then sat back with satisfaction.

"I thought so!" he said. "Says here that is you qualify for one of these disaster loans, then our agency is *obligated* to loan you your production money for not one, but the next *two* years."

"That's a lot of sugar for a nickel," I said dubiously.

"Well, that's what it says," he declared. "Here, fill out these forms." I did so.

"Now, you don't want me to go to the other lender and get a subordinated loan, or a participating loan, or anything else?"

"Nope, we got you taken care of," he told me patiently. "All you have to do now is wait on the paperwork."

The one thing I did right at this point, as I left the

office, was to make a commitment to myself not to put a seed in the ground until I actually had the money in the bank. Since this was the second week in March and I had rented out the cotton acreage, I felt safe in making this promise. After all, I could wait until June to plant beans, if I had to.

Things rocked along for two months. The renter and buyer both planted their crops and got a stand. Their money was still on the table. My loans were delinquent and getting more so. Lender one still would not give a release or a non-disturbance clause. Then the end of the first week in May, I got a short hand-written note from the disaster loan man, lender number two. It said simply, "Under new regulations, we can no longer make production loans through the disaster program. Your disaster loan is approved, but you must arrange for your production money elsewhere."

Remember, I had been waiting two months under the impression that this money was a lock-in. There's a lot of difference in approaching a bank for a furnish loan in March than in May. I finally got a subordinated loan worked out with Lender one and notified Lender two to cut the check. All the guy ever said about the two months waiting was, "Sorry."

The disaster loan check came in on June 4. No one called. No one wrote. I found out about it second-hand on June 12. I had even been in the office twice during that period and no one had said a word to me about it. I called Lender two. "I hear my check is in," I said.

"Sure is. But we'll have to hurry to finish up the

paperwork; those things have to be cashed in ten days or they go back." No answer as to why someone had not called.

I spent two days filling out more forms for a "no-contingency" loan. Finally I had the check, plus the deposit slip filled out in my hand. The man slid one last piece of paper across the table for me to sign.

But I read it.

"Hey, there's a mistake here. My house is on this mortgage!" I pointed out.

"That's right. We have to have a mortgage on your house," he declared.

"You never mentioned this three months ago!"

"Well, we still have to have it," he shrugged.

"Where'd you get the description?" I asked.

"Down at the courthouse," he replied.

"Okay, let me see if I understand. You'll make me this loan with three hundred acres of farmland and the house as collateral?"

"That's right."

"Well, there's three hundred more acres of farmland I own you must not have picked up off the land records. What if I put it on there and we take the house off?" I thought I had him here.

"Nope. But I'll take that three hundred acres, too," he said calmly. We argued about that for a while.

"Okay," I allowed, "let me see if I understand now—six hundred acres plus the house is enough collateral for this loan?"

"Yep."

"Well," I knew I had him now, "there's three hundred *more* acres you didn't pick up at the courthouse. How about I put that on here and we take off the house? That'll be nine hundred acres total."

"Listen. I don't believe you understand yet. If we make you this loan, we will have your house on this note!" The man was very firm.

So was I. I walked away and left the check on his desk, knowing there would be bad consequences, but also knowing I'd be a fool to bet my house against a soybean crop planted the third week in June. As it turned out, it didn't rain again until late August. A neighbor who planted his beans that same week told me he combined seventy bushels off of one hundred acres. Seven-tenths of a bushel to the acre!

Less than a month later a real estate agent came by with an offer to buy about 530 acres. The rent money and the thirty-acre sale money were still on the table in July. I signed an option and went to see Lender one about working it out. For one hour I met with two of their officers. They were very favorable; matter of fact, one of them even said, "If this works out, your loan will be in a preferred position." The option was a sixty-day agreement, to enable us to try to work it out with all the lenders. Therefore, I assumed things were okay for at least another sixty days.

Remember the old saw about "assume"? It makes an "ass" out of "u" and "me"!

One of the Lender one men told the other, "Let's go ahead and get that $75,000 off the table. Execute the rent and thirty-acre sale agreements." The junior man turned to me.

"Tell the lawyer to go ahead on those two agreements," he instructed.

"Nope," I demurred firmly but gently, "I have told him four times in nine months to go ahead with those contracts, and every time ya'll backed off. If you want it to go through, pick up your phone and call him." To my surprise, he did just that, instructing the attorney to try to get it done by the end of the week.

As I left his office, he walked down the hall with me. In the presence of three bank employees and two customers, he put his arm on my shoulder and assured me, "Don't you worry about a thing. I'm going to take care of you!"

Good feeling, right?

Within forty-eight hours, without another word or even a written note to me, this lender started foreclosure.

Well, it gets sticky from here on, complete with lawyers and all the trimmings, so we'll skip all the gory details. If I could just understand it! At this date, the rent and thirty-acre sale money still have not been executed, and both men almost have their crops out. The buyer of the thirty acres says he's still willing to buy, but "not for nine hundred dollars. Maybe seven hundred and fifty dollars." I don't blame him. Had he been able to buy a year ago, he would have put in a well for his crops. They burned this past summer without irrigation, so both of us took a loss. Even with irrigation, the renter still took a lick from the drought and heat, and says he may not can come up with all the rent. But he was willing to pay it in advance a year ago.

Why? You got me!

All the chips haven't fallen where they may as yet. I've been to Congressmen, Senators, lawyers, bankers, family, and friends to try to salvage what I can, or at least try to *understand.* There's a real need for someone who does understand those situations to try to act as an arbitrator, an intermediary. Maybe I'll try that. I'll let you know. If Dr. Jim doesn't have me committed before we get this one into print, I was hoping to get him in on a sequel—entitled: *Is There Life After Farming?*

Take it, Jim.

Good luck, guys.

BOOK TWO

Coping with the Loss

Chapter 1
The Problem

I begin this chapter by describing my credentials to write about the farm crisis. By some measures these credentials are weak. I have been working closely with the farm community for only a couple of years. I am not a farmer. I have not extensively researched the "farm problem," although I have read several research papers. The main sources of my information are the individual farmers and their families, with whom I have worked in private counseling, and from participants' discussions at seminars I have presented to farm communities (nine to date). I have had some formal training through the Minnesota Extension Service in farm credit mediation. Using this information I am assisting the Mississippi Department of Agriculture in developing a training program to mediate between farmers and lenders.

In the fall of 1984, I was referred several farm families for counseling. They were stressed by the financial burden of farming, which seemed to be getting worse.

One year later I was invited by the Episcopal church in Rolling fork, Mississippi to talk to local farmers on "stress management." I arrived at the church a little late, and the audience was already beginning to arrive. The parish hall was eventually packed, standing room only. Not only farmers, but bankers, merchants, a doctor and hospital personnel, and ministers from other churches were there. I listened to joking and nervous laughter before the start of the seminar. I found out later that farmers don't like to admit that they may need help with stress. Most passed off the seminar as "the only entertainment in town."

Of the nine seminars, this first one was the best attended. Foreclosures were on the horizon, but not yet a fact of life.

As the problem worsened, fewer farmers attended. Concerned wives, friends, and interested others—including a handful of farmers—came to "find out what they could do to ease the pain of those going under" and perhaps prevent a suicide. At every seminar I was told, "Old So and So should have been here. He's the one that really needs this." The needy ones seldom showed. They were already isolating themselves. I began to realize that seminars were not the way to minister to farmers in crisis.

The crisis in American farming is real. The losses are real, in terms not only of land but also of human suffering. There are losses of self-respect, self-esteem and self-confidence. Faith in God is often lost along with a loss of hope for the future. There may be a loss of obligation to generations past to keep the family farm intact, a loss of the children's education or a reduced concept of their education, a loss of standing in the community. And the list goes on.

My strongest credential is an understanding of the process of losing something of great value and of the ways to cope with that loss. My interest in collaborating on this book is two-fold: we may be a witness during our time, not only to a few farmers going out of business, but to a drastic change in the institution of farming in America as we have known it. Some of the reasons for the change need to be documented for history. Secondly, those victimized by the process of change need skills to cope with it.

After reading Bob's story, you don't have to be a psychologist to know that he is angry. Some readers may discount his credibility, others may empathize. I do know that Bob's case is not isolated, or even very unusual. As many as fifty percent of the farmers may leave the land, in some areas of Mississippi, victims of an economic system they had depended upon.

The following examples of farmers' complaints (a partial list) come from the *AG-Link* hot line for Oklahoma farmers, published in December of 1985:

1. Lending institutions make agreements with borrowers but will not put agreement in writing.
2. At a time frame convenient to lender, lender will pressure farmer to liquidate disregarding and violating verbal agreement to farmer.

3. Lending agencies appear to be "looking for a bargain" to foreclose.

4. After a farmer loses farm through various procedures, lending institution will let another person purchase same farm for much, much less money and less than half the rate of interest.

5. Lending institutions tell farmers they will help, but when farmers go in to agency the lender talks "liquidation."

6. Lending institution reprimands farmer for creating unplanned-for additional expense when farmer had heart attack.

7. Lending institutions tell farmers to purchase seed, fertilizer, etc., and agree to cover cost to dealer. Farmers make purchase and debt, then institutions back out, leaving farmer "holding the bag," resulting in possible/probable foreclosure.

8. Farmers indebted to more than one lending institution and being current in all except one are being forced out by the one. The farmers cannot secure financial assistance to help them over the hurdle with the one lender.

9. The farmers relate similar statements coming from borrowers of all institutions that the final word is "pay up," or "get out," rather than work with the farmer.

10. Farmers who have done business all their lives at a particular bank have in several instances learned bank failed. New people come in and refuse to honor agreements made with farmers prior to bank failure. Foreclosure.

11. FmHA promises help to farmers, who pass this information along to creditors. Farmers straighten up ill-feeling and business with creditors, but when

creditors ask, the FmHA tells creditors a completely opposite story.

12. FmHA will work up loans for farmers and let them remain untouched for months, resulting in loss and sale of land at sheriff's sale. Farmers are willing to make public statements to this affect.

13. All lending institutions demand unmortgaged property farmer has or refuse assistance or initiate foreclosure or involuntary or voluntary liquidation.

14. Lending institutions refuse to work with farmers on partial-payment plan to satisfy more than one lender. Each lender wants ALL, resulting in inevitable foreclosure.

15. Farmers who have never before been late or delinquent in payments are being told if payments are not kept current or are not on time, institution will commence foreclosure proceedings within thirty days. No negotiations are accepted.

16. Banks that have lost money in oil transaction loans foreclose on farmers who are in "good shape," except for debt/asset decline ratio.

17. Lending institutions "devaluate" machinery and collateral on farmer, causing debt/asset ratio to fall, resulting in foreclosures.

18. Institutions initiate foreclosures against farmers even with interest paid current.

These statements above don't necessarily prove anything except that many farmers report similar experiences with lenders and that Bob's problems are widespread among farmers. From the lender's point of view he has lent his money in good faith and expects to be paid back. (I have worked with lenders who are fully as stressed as the farmer.) When the farmers are in a crisis, so is the farm credit system.

Effects on Farming Communities

The failure of a farm has a ripple effect on the community at large—even on some who seem uninterested in what happens to farmers. Kathy Mangum of the Minnesota mediation program has classified the impact of the farm financial crisis according to three categories: first order effects such as tax and loan delinquencies, hunger, stress, lapse in medical insurance; second order effects such as relief and volunteer efforts, anger and scapegoating, retail business failures; and third order effects such as county and municipal budget crisis, closing of smaller schools, population losses.

The first order effects are felt by the persons directly involved—the farmers, their families and to a lesser extent, the lenders. The second order effects are effects on the community, and third order effects reach to the local institutions and government.

Enough about the causes of the farm crisis. My job is to identify the emotional problems which result from the crisis and to outline healing strategies. The story is pretty much the same from farmer to farmer, although the degree of the problems experienced may differ. The sequence of events and emotional consequences are as follows:

1. The farmer becomes frustrated by various blocks

to doing his job and paying his debts. The government, the weather, lenders seem to form a conspiracy to harrass the farmer. The farmer reacts with anger and complaints. Some hope remains, however, since he has faced similar frustrations before and found relief "next year."

2. The first signs of powerlessness begin to occur. The farmer discovers he will be in trouble no matter what he does. The anger escalates, sometimes reaching rageful proportions. Family members (wife, brothers and sisters with an interest in the farm) may remember his decision to buy land or "spend big" during successful years, and remind him, "if only you hadn't. . . ." He feels the first signs of isolation from the family. There is a resentment toward self as well as toward others.

3. Anxiety and worry cause sweaty palms and sleepless nights. Lack of sleep results in a short temper the next day. Anger feels better than the anxiety, and blame becomes a standard when farmers gather together. When alone the farmer may harrass himself about letting down his family, his creditors, and his ancestors. Drinking, drugs, denial, excuses, and blame are all excapes from these bad feelings. The worst is the sinking, weak, sick-at-the-stomach, suffocating feeling each time the farmer faces the powerlessness to prevent the loss of some of the most precious things in his life. He feels like a victim and struggles to free himself of that role. It is not unusual for physical symptoms of

stress to appear; ulcers and other gastrointestinal problems, chronic headaches, impotence, heart attacks, increased frequency of colds and flu, to name a few.

4. Isolation and loneliness add to the farmer's stress. Farmers believe, often falsely, that "nobody wants to be around a loser." They don't like to attend a party with the chemical dealer, an old friend whom they owe and will not be able to pay. (One of my workshops was organized by an agri-businessman who would lose many thousands of dollars but still maintained compassion for his farmer neighbors and friends.) However, the farmer may project his own anger toward himself onto the hearts of others. Sometimes friends and family do show signs of rejection. In any case, isolation is a frequent result. One recent study showed that sixty-two percent of the men and sixty-six percent of the women of farm families in economic crisis withdrew from family and friends.

5. Lonely already, the farmer separates himself from God. He feels that God has let him down and often reports being angry with God. "Why me?" is a frequent thought. Most of all he is mad at himself. With isolation from family, social supports and God, he feels worthless. The farmer is now depressed, feeling hopeless. (The 1985 study mentioned above reported ninety-seven percent of the men and one hundred percent of the women reported depression. Fifty-six percent of the men and seventy-two percent of the women continued to feel depressed over time. The higher percentages in the beginning probably included the nor-

mal grief reaction and the later figures are more representative of a clinical depression.)

6. A small number of these farmers will feel desperate. Persons in despair behave in desperate ways. Some consider suicide or murder as a final desperate act, and some will actually follow through. Suicide is an act not always provable. Hunting accidents, automobile accidents, and other "careless acts" can result in deaths, the facts of which are often unclear.

Some farmers don't progress through all of the steps. They may accept their losses and start life over. In the study above, twenty-eight percent of the forty farmers studied moved away and eventually obtained jobs at a professional level. Some returned to college or other retraining programs.

I have been asked, "Is there a 'farmer's personality'?" The answer is no. Farmers may be aggressive or passive, outgoing or reserved, calm or nervous. They come in all sizes, shapes, and flavors. They do, however, have some common characteristics. They are for the most part independent and like to do things their own way. They want to be strong, covering up any hint of weakness or need for help. They are proud. These characteristics make them a difficult population to help.

The pronoun "he" is used to refer to the farmer. In fact, all of the farmers with whom I have personally worked have been male. However, the farmer's wife has had a significant role in these troubled families. Usually the woman has as a primary goal holding the family together, no matter what happens financially.

Other than in this generality, farmers' wives vary widely from each other. A few wives I've worked with were more stressed than their husbands. The relative helplessness of their position frustrated them greatly. When their husbands procrastinated on business or made poor judgements, they looked on inadequately. Some farm wives were either the major decision maker or shared equally the decisions with their husbands. The Mississippi Delta is still chauvinistic, and these more aggressive wives reported being ignored by lenders in a financial conference. They were forced to communicate indirectly, through their husbands, much to the discomfort of all involved. Very often the wife was the source of strength that kept the family going.

Farmers and their families who have lost the family farm need to be trained to avoid remaining chronic victims of the crisis. They need special skills in: (1) recovering from the victim experience, (2) putting their losses behind them, (3) regaining a sense of self-worth, and (4) and coping with the conflict. The following chapters will address these skills.

CHAPTER 2

The Victim Experience

Bob was a victim of the farm crisis. He was caught in a financial crunch over which he was powerless. An important distinction for you to understand is the difference between a *crisis* and a *chronic* victim. Bob was a crisis victim. He may or may not become a chronic victim, depending on a number of factors which you will learn about in this chapter.

When you begin to sense the possibility of losing something—be it pride, control, influence, money or whatever—you will take actions to prevent that loss. You may, for example, attempt to solve the problem, usually by changing your behavior or your approach to the situation. You may want to persuade someone else to change or to compromise with a competitor. If the threat of loss continues, you will eventually begin to feel like a *victim,* changing your approach to the loss situation.

Victim Response

Victims will aggressively attempt to control the emotions of other people involved or they will passively attempt to make someone else responsible to "fix things," or they will escape. If you are a victim, the goal of your behaviors, then, is to dominate the emotions of others to get them, against their will, to do your bidding, or to escape the pain of the loss.

AGGRESSIVE VICTIM BEHAVIORS

The belief that "If I make others feel badly enough, they will do it my way," is the credo of the aggressive victim. You may be thinking, "That's not a victim. That's the kind of person who makes victims out of other people." You're right! However, victims are victimized by other victims. You would not want someone else to suffer unless you felt like a victim or potential victim yourself. The aggressive victim is motivated by anger. When you take on this role you will attack, criticize, blame, intimidate, coerce, and perform any other behavior that might domintate someone else's emotions in a negative manner.

PASSIVE VICTIM BEHAVIORS

"I'll suffer and make others responsible for me. I know if I suffer enough someone else will do something for me. If no one else, maybe God will take pity on me," are beliefs which motivate passive victim behaviors. The passive victim prefers behaviors that communicate suffering: being hurt, upset, afraid, depressed, dependent, or feeling self pity. The passive role is the "classic" victim role. When you act as a passive victim, you hurt and want others to know it, so one of them will take care of your pain.

ESCAPES

In spite of your best efforts, other people may successfully resist being manipulated to do things your way or to take care of your problem. The next line of defense is to escape. Escapes can be grouped into three major areas: physically getting away, psychologically getting away and chemically getting away.

Physical escapes involve running away from the area or avoiding the situation. Rarely this may be okay. However, there are very few problems that don't move with you or that aren't still around when you return. Some farmers have spent more time away from farm business as their problems increased.

Psychological escapes are the most frequent choices. You can't always physically get away, but there is always a handy psychological escape. These are "mental" ways of avoiding the problem. For example, you can deny that there is a problem. We all are capable of denying the existence of the most obvious problems. You may prefer to distract yourself with excessive work, overeating, or other compulsive activity. You may distort reality, pretending that the problem is someone else's to solve. You may minimize the problem as such a "little thing" that it doesn't deserve your attention. You may blow your own idea of yourself up to such proportions that you are above problems and can't be hurt by them. You can emotionally withdraw from other people, even when you're in their company. You can block your emotions so that you feel only certain kinds of feelings or no feelings at all. Escapes give temporary relief at best and over the long term can create psychological symptoms more painful than dealing with the hurt you're escaping.

Chemical escapes include alcohol and other drugs. You may escape the pain of conflict or loss by dulling your brain. The longterm difficulties of this method of escape are among man's biggest problems to solve.

SURVIVAL

These victim behaviors are part of our survival system. All human beings rely on these behaviors at certain times and under certain conditions. During our primitive times it was a survival matter to get physical dominance over threats, but physical dominance is not always acceptable in today's world. We do permit emotional dominance and control. However, people in the same environment seeking to control one another's emotions account for much of the pain in this world. These survival behaviors may create more problems than they solve. The fact is, it is *you* who determines the situations when these victim behaviors will be used.

John was verbally attacked by Bill, his neighbor, after his dog relieved himself on Bill's lawn and Bill stepped in it. John was angry, and he wanted Bill to feel bad in retaliation. He fought back by bringing up all the complaints he had stored in the years of being Bill's neighbor, focusing on Bill's unruly children.

Alice was not invited to a party at her friend Judy's home. She thought that Judy was one of her best friends. By the end of the next day, having thought about it continuously, she was furious. She called Judy and unleased all the complaints she had stored over the years of the friendship including Judy's unruly children.

George decided to break his dependency on his family. Instead of calling home several times a week, he decided to call every two weeks or so. Before he made his intended call, his mother called him and unleased all the bad feelings she had stored.

The point is, under invitations of various strengths we may feel we have been victimized. Whether or not you are a "victim" is a personalized decision or perception rather than a fact. One measure of emotional maturity is to use the victim behaviors less and less, as a response to threat. When you choose to attack or suffer—to manipulate other people—you invite retaliation or rejection. When you choose to escape, problems don't get solved. Victim behaviors have a low probability of satisfying needs and should be used sparingly or dropped quickly when recognized.

BLAME

Blame is the most damaging of the victim behaviors. It is the only behavior that is aggressive, passive and an escape. It is sometimes meant to make others feel bad.

Other times blame may be an invitation for others to do something for you; after all, it wasn't your fault. Finally, blame is an escape from responsibility to solve the problem. The process of blame often fools the blamer. Once you have searched for and found the fault and then blamed the guilty, you may feel as if you have actually done something constructive. In truth, however, nothing has changed! Finding the problem is only the beginning of problem solving. Blaming can be an end in itself. If you ended Book One with a feeling of satisfaction that Bob has identified some of the "bad guys" (outside forces that contributed to the farm problem), you have done little other than give yourself temporary relief. The problem has not been changed by this understanding. Knowing causes is a foundation for problem solving, not a stopping place, unless you expect someone else to solve the problem for you—good luck.

Effective Responses To Conflict

What are the alternatives to behaving like a victim? First of all, victim or survival behaviors are our most natural response, though not always appropriate or effective. To be more effective you will have to override your natural feelings in the beginning. Of course you have already done some of this by simply growing away from childhood. Children think everything is a "big deal." As an adult you have already defined many things that upset you previously as "no big deal." If there is no significant loss or no threat, this action of making it a "little thing" is the most efficient.

A good analogy of this action is the family dog. He is sleeping peacefully when an unfamiliar person enters the house. He becomes alert, listens for sounds, looks for signs of danger and sniffs the air. He may even go over to the intruder and sniff at closer range. When his "dog computer" registers "no real threat," he settles back into his warm spot.

In the case of Alice above she might have defined the lack of a party invitation as no significant loss to her. If one of you should write me a letter saying that this book stinks, I might get a bad feeling in my stomach, but if I rethink the situation, put it in perspective, I will realize that I have not suffered a significant loss. At the most I have lost some false pride. I would be grandiose to think that every reader would like or benefit from my part of this book. Your disagreement is okay. I may even contact you for feedback on the parts that troubled you. Your letter may turn out to be an asset. Now if my publisher writes that he doesn't like the book—that is a significant loss and it would be out of reality to think, "No big deal."

Other things also are a "big deal." If your father is considering disinheriting you or your spouse has asked for a divorce or your lender keeps you waiting past a reasonable planting time without a yes or no decision on your loan, important things are happening. Lesser things are also a "big deal." Thinking again about Alice's not being invited to the party, Judy's rejection may signal a problem in her relationship. A partner or family member may be putting an unfair burden on you by not doing his fair share of the work. A repairman may have done a sloppy job. A politician you supported may vote against a bill that would help you. These conditions all contain problems to solve.

Problem solving means that you *think* about what's going on. The victim behaviors discussed earlier don't require thinking. They are feeling-based and occur without much thought. Thinking includes figuring out exactly what the problem is and what outcome you are looking for once the problem is solved. In other words, you need to clarify for yourself what is going on now that you don't like and how you want conditions to change so you will be happier. The final steps involve: (1) thinking of an action which might bring about some part or all of the desired outcome; (2) taking a new action; (3) evaluating your progress, and (4) deciding that your problem is solved or continuing to think up and implement new actions.

All of the above actions (victim behaviors, redefining, and problem solving) are motivated by the same feelings, usually anger or fear. You may think, "I wasn't angry or afraid, I was only "frustrated" or "cautious." These two terms are forms of anger (frustration) and fear (caution) at lower levels of intensity. All are natural responses of human beings. The victim behaviors are the most natural (most primitive) and least effective. The redefining as "no big deal" takes the least effort when it applies. Problem solving requires the most conscious effort but has the largest potential payoff. The results of problem solving go beyond solving that one problem. There is often change or "personal growth" within the problem solver. His or her confidence is increased with each successful solution.

Hints to Avoid Becoming A Victim

1. Don't ask other people to do things for you that you could do yourself.
2. Don't expect others to think for you.
3. Don't feel sorry for yourself—do something constructive instead.
4. Don't believe other people are making you feel bad without your cooperation.
5. Don't expect others to make you feel good.
6. Don't help others unless you are sure they want you to and you sincerely want to.
7. Don't take responsibility for the way others feel.
8. Don't get between other people in conflict.
9. Don't blame.
10. Don't take responsibilty to punish for the sake of justice.
11. Don't seek revenge.

Exercises

1. Think of the conflicts, threats, or stresses—big or small—that have happened to you in the recent past.

2. Determine in how many of those situations you behaved like a victim. In those situations did you get what you wanted? Even if the victim behaviors seemed to work out for you, remind yourself of the low probability of these behaviors meeting your needs over time.

3. In the future, increase your skills in identifying

those victim responses. Remember, victim behaviors usually create more problems than they solve. The first step in changing to more mature responses is to recognize the problem behaviors.

Chronic Victim

When you began to see the "handwriting on the wall" that you are going to lose, your feelings will intensify. You will experience yourself in a crisis. To protect yourself from experiencing the loss, you will begin to deny that it is happening. Denial is powerful, and otherwise sane persons can even deny for a time the death of a loved-one. One part of denial is to physically deny feelings. You know that it happened but you feel "numbed out" instead of being naturally sad. In the extreme, denial may take the form of depersonalization. You may not know who you are, or you may feel as if you are outside of yourself.

Fear is usually present. There may be waves of fear, sometimes growing into panic. Fear may alternate with anger. When you get mad you will feel less fear. Anger may intensify into rage. Anger usually feels better than fear since there is a hope with anger that one can "get even" or that justice will be done.

Magical kinds of thinking may occur. If you are just "good enough" or refuse to accept what is happening or bargain with God just right, then the problem will go away.

Most persons will, in time, accept their losses. Some

don't, and those who don't continue to deny the loss, stay resentful, and magically hope for relief. Though resistance to accepting loss is natural for a time, it is inappropriate and self defeating to resist when nothing else can be done except stay mad, deny, and hope. Those who never accept the loss become *chronic victims.* Each time they are reminded of the loss they return to one or more of the victim behaviors. These chronic victim experiences are cumulative and result in a person's chronically behaving like a victim under most circumstances. Any of us may be a *crisis victim,* but we all have a choice as to whether or not we become *chronic victims.*

CHAPTER 3

Turning Resentment Into Grief

Anyone may suffer losses through illness, accident, financial crisis, death, unfairness, natural disaster or other tragedy. Some of these unfortunate individuals never fully recover. They carry a little piece of the crisis forever. A little hurt here, a little resentment there, and some "It's not fair" are stockpiled. If these losses are not dealt with, they will accumulate into a chronic victim experience. Chronic victims never quite get over their past. Other persons are able to return to normal functioning following the worst of tragedies. The process by which they return to normality is *grief.* Grief is the process of letting go of whatever was lost.

In her book, *On Death and Dying,* Elizabeth Kubler-Ross outlined five stages of grief: denial and isolation, anger, bargaining, depression, and acceptance. Kubler-Ross was writing about patients who were facing their own death, but her findings apply to persons facing any major loss. She noticed that the first response was to deny the information. The dying patients would delude themselves that the x-rays or lab tests had gotten mixed up. Kubler-Ross states, "Denial functions as a buffer after unexpected shocking news, allows the patient to collect himself and, with time, mobilize other, less radical defenses." Hospital staff and family members may be so uncomfortable with this distortion of reality that they begin to avoid the patient.

When the patient faces the truth, the question "Why me?" often follows. Anger replaces denial and these feelings may be expressed toward anyone who happens to be around. The anger brings no relief so the person moves into another tack—to ask God nicely for help. "If I do X, maybe God will let me live" is the basis of this bargaining.

When the person can no longer escape facing death, depression sets in. The person falls into deep despair facing the loss of everything that continued life would offer. Depression prepares the way for acceptance. After the pain of depression comes acceptance of the inevitable death.

Losing the family farm initiates the same emotional factors as facing a death. For your understanding I want to split reaction to loss into two major categories—first, preventing or escaping the loss and second, dealing with the loss. Denial, anger, and bargaining are all designed to avoid the loss. Depression and acceptance are characteristic of dealing with the loss.

Powerlessness

The capacity to cope with being powerless to prevent a significant loss is an important human skill. Life is a series of losses. As a child we must deal with friends moving out of the neighborhood, pets dying, learning the truth about Santa Claus, disappointing our parents and thousands of other losses. These losses all result in the grief process. If this process is completed—that is, if we move from denial all the way through acceptance—the loss is "put behind" us. If the process is not completed, the loss continues to affect us in some manner. A person who holds on to past hurts will still be resentful, think of the injustice, seek revenge or the return of justice. Holding on to the past can result in psychosomatic illness, emotional pain, and disturbed relationships. It is a major characteristic of the chronic victim.

Changing Resentment Into Grief

In order to simplify your understanding, I am going to label moving through all the steps of Kubler-Ross's grief process as "grief." If acceptance is not reached, I will call this desire to avoid the loss "resentment." So grief includes the emotional pain (depression) of accepting and letting go of whatever was lost and resentment includes denial, anger, and magical thinking designed to hang on to whatever was lost. The major theme of this part of the book is coping with loss. *Successful coping means that the person is able to give up resentment and to grieve instead.*

Why wouldn't everyone naturally move through the resentment into grief? Isn't grief the natural thing to do? Grief is natural, but not everyone grieves. Plenty of people have no history of grieving. Grief means experiencing the pain of losing. If you don't feel anything or feel only anger, grief will not be completed. Even if you are involved in heavy suffering with resentment as a part of the suffering, grief will not be accomplished. That kind of suffering can go on forever. The anger in resentment motivates overcoming something, not giving in and accepting. A rule of thumb is: *as long as you stay mad, you won't completely grieve.*

Resentment is not easy to shed. Our culture values resenting whatever seems unjust. To give up resentment when you have been "wronged" may seem as if you are "wimping out." There is a magical belief that if you just hold on to the resentment, justice may possibly be done. Further, according to the magical thinking, if you give up the resentment, you give up your chance to right the wrong. As long as you stay mad, you won't have to accept that you are powerless. You feel strong, but you are actually stuck, unable to put the loss behind you.

IF YOU CAN'T CRY

"I can't cry even when I want to," I have heard many potential grievers complain. Although crying is natural, you can block the emotion. You may have blocked your own tears by biting your lip, tightening your stomach, or any one of a number of other distractions. In the steps that follow you will be asked to consider your losses one at a time. If you hold the thought of a loss in your mind and concentrate on the area behind your eyes, tears may flow. If you add deep breathing, filling up and then emptying all the air out of your lungs, the probability of crying increases. Keep up the deep breathing for five to ten minutes if necessary. Be sure you are sitting down because you will probably get dizzy. If you are pregnant or have a heart condition or other physical problem, you need to check with your doctor before using this technique.

Learning to Grieve

The following steps will help you move from resentment through the grief process and put the loss behind you:

1. *You must first accept that you are powerless to do anything constructive about the loss.* That is, you must accept the hopelessness of the situation. Giving up hope is painful. If there is real hope, of course, grief (letting go of what has been lost) is inappropriate. If you can actually take a corrective action, you should be solving the problem, not grieving.

2. *You must be willing to give up the resentment.* If you don't feel right or ready to give up your anger, you have a right to keep the resentment as long as you wish. Some people use the energy from their resentment as motivation to achieve. Some honestly feel that remaining resentful is only right and may feel that being resentful is magically getting back at someone. You may need to suffer more before you're ready to give it up. If you only half-heartedly decide to give up the resentment and continue with the following steps, it will likely come back anyway.

3. *Next make an exhaustive list of your losses.* Don't simplify or summarize your losses. Instead search extensively for all the losses resulting from your experience. The following losses were collected from victims of the farm crisis: land, family heritage, income, borrowing power, something to leave to my children, my children's education, friends, comfort, a belief in God, strength of character, the ability to forgive, energy, motivation, happy times, security, perspective, a belief in the future, a belief in this country, a belief in justice, a belief that honesty and hard work are enough, self-respect, pride, self-confidence, family respect, family relationships, farm equipment, the belief that everything happens for the best, a belief in justice, a belief that other people care, a sense of logic, a sense of reality, my sanity, time, the ability to care, peace, durability, determination, a sense of accomplishment, looking forward to another day, productivity, power, cause and effect, toughness, control of my life, mas-

culinity, desire, eagerness, choice, leisure, boldness, excitement, enthusiasm, hospitality, humor, sobriety, and charm.

An important loss sometimes overlooked is resentment itself. Giving up the resentment also needs to be grieved over when giving it up is experienced as a loss.

4. *You must separate the cause of the loss from the loss itself.* When you think of the behavior or situation creating the loss, anger is a frequent reaction and stops the grief process. Think of the losses on your list, one at a time, and experience the pain of grief for each. Avoid thinking of the cause of the loss and the risk of getting mad. If, at the end of grieving for all the losses you still feel strong resentment, you have probably missed a key loss. For example, a woman grieved over the loss of her boyfriend, but she still continued to resent him. The key to her block was her unwillingness to give up the fantasy of "his being sorry someday."

5. *Finally, develop some symbolic representation of separation from both the resentment and the losses.* The symbol may be an act or a fantasy. Some choose to write down resentments and losses and bury them in the back yard. One person, while on vacation, let go of the list, weighted with a rock, from a cruise ship. Fantasies of actions such as these are also acceptable.

CASE HISTORY

Ben thought of himself as a decent, law abiding, fair person, although his recent behavior suggested the contrary. He had always been a fighter in life. He hated to lose, whether an argument or a tennis game. He had experienced minor difficulties with authority figures in the past. He believed in doing things "his way."

A few weeks before he sought counseling, Ben had set up the murder of his brother. Ben waited in his car, out of sight, but where he could watch for his brother's car as he returned home. It was 7:30 P.M., and he could expect his brother to appear around 8:00 P.M. He sat in the darkness, drinking from a fifth of vodka. In his lap was a .38 police special. Out of boredom, he briefly turned on the interior light and penciled a list of names. His brother's name was at the top and his own name finished the list. In between were the names of five other family members—it was a death list. Although he did not want to die himself, he was not willing to face the consequence of a multiple murder. It was getting cold and he was afraid the heater would discharge the battery, and he began to drink in big gulps to ward off the chill.

His brother didn't return at the expected time. Ben waited and continued to drink. He awoke at 3:45 A.M., still drunk but his hate dulled. He had no stomach for killing. He was cold and his body racked with pain from sleeping in a cramped position. He drove home feeling totally defeated.

Two years before, Ben's remaining parent, his mother, had died. Since she left no will, the heirs had the task of deciding how to divide the assets, including a farm that had been in the family since the mid 1800s.

Ben loved the land, and he assumed his family would agree to become partners with him in the farm. He was the only one actively involved in running the farm. Fifteen years earlier, shortly before his father's death, he had promised him that he would keep the farm in the family at all costs. Now, however, his brothers and sisters wanted to sell the farm and split the money, and Ben couldn't raise the money to buy the land himself.

His first ploy was to threaten the family with the public disgrace of a court fight. They were disappointed but not moved from their position. Next he withdrew and left vague threats of suicide. As Ben faced his powerlessness, he began to feel panic. He would not accept the loss of the farm. He next felt "numbed out" for a short time. He replayed his conversations with his family, searching for some hope on which to build. His resentment increased with time. He hated his family. Although not an alcoholic, Ben began to drink excessively every night to control his anger. The drinking had an opposite effect. As he drank, he plotted revenge.

Six months later he was in a depression. The farm was on the market but not yet sold. He was drinking very little now. His family doctor had given him "pills" for his depression and advised him not to drink with the medication.

Ben began to obsess about life being meaningless if he lost the farm. He had broken his promise to his dying father. He had lost his means of supporting his family. To add to his problems, his wife had left him and would probably file for a divorce. Ben thought of suicide or running away, but the fantasy of shooting his family members who had disrupted his life was more satisfying. He felt the power of the fantasy for several months before deciding to act.

In his treatment, Ben could accept that he was powerless. The therapy was slowed, however, by his unwillingness to give up resentment. Even more painful than losing the farm, at this point, was his mental picture of his brother smiling after convincing the family to side with him against Ben.

After two months of grief work, Ben was no longer depressed. He could think about losing the farm without resentment. He had accepted that loss. He still was angry at his brother. He chose to avoid him, but he no longer wanted to kill him. His most intense grief was over his mother's failure to make a will, but he now had forgiven her.

Some farmers with whom I have worked and talked have carried hate in their hearts for those who shape government policy, lenders, family members who have let them down, former friends, local merchants and often for themselves. In listening to their reasons I find their arguments to be sound. They have good reasons to be angry.

You may have good, defendable, rational reasons for your personal resentments. *You have a right to suffer, but you don't have to.* My hope for you is that you will decide to accept your losses and put them behind you.

Chapter 4
Regaining Self-Worth

The experience of losing the farm will usually affect the way the farmer feels about himself. Sometimes the effect is positive. Facing adversity and surviving can build character. Families may become closer through facing the crisis together. Self-respect is a frequent by-product of courageous behavior during the misery of total defeat. Turning the defeat into a new beginning builds self-confidence.

The flip-side of gaining personal worth through surviving a catastrophe is losing self-worth through a sense of failure. The failure is personalized and equated with being worthless. How can two farmers, suffering similar misfortunes, have two opposite consequences—one gaining through loss and the other losing self along with substance? The answer lies in the manner in which each farmer determines his own value.

Self-Appraisal

Your personal worth is determined through a combination of self-appraisals. As a child, most of you got both conditional approval and unconditional love from your parents. A lesser number received only conditional approval for things you did, and very little unconditional love for what you were. A very few of you may have felt disapproval, no matter what you did. This approval was an important part of your development into an adult.

Without giving it much thought, you probably assess your value as a person through the following perceptions of personal experience: (1) If you were treated as "lovable" by your parents, even after making a mistake, then you will continue to love yourself unconditionally (you don't have to earn it), even when you make mistakes today. You can also earn a sense of worth by: (2) developing personal skills in accomplishing goals and communicating with others, (3) getting the approval of others, (4) self-glorification (building illusions of being "better than"), (5) self-discipline, and (6) problem solving skills. These six appraisals collectively determine your overall sense of personal worth.

UNCONDITIONAL SELF-LOVE

Children who have a history of being socialized by their parents without withdrawal of love, grow into adults who have a basic sense of worth even when they are not earning it. That is, they continue to love in good times and bad times. Those of you who were loved by your parents on a more conditional basis, may feel that it is wrong to love yourself when you're messing up.

For example, Greg decided to expand his farm. He bought the land, and within two years the bottom dropped out of land prices. Greg felt worthless. He found it difficult to face his mother who still owned part of the farm operation. Was Greg really worth less as a result of a shift in the economy? Would he have been a better person if land prices had risen instead? If Greg had the capacity to love himself unconditionally, he would have grieved over the loss of land value, but he would not have lost any self-worth. He was neither better nor worse as the result of fluctuating land values. He was the same person, just less wealthy.

APPROVAL OF OTHERS

Some of you may continue to need direct approval from other people as a "steady diet" to feel okay. You may not easily love or approve of yourself, and you don't fully trust approval when you get it. Oh, it feels good, but the effects are short lived. Approval becomes a major guide in your life—you do those things that others approve and avoid those actions likely to bring disapproval. The problem is that no matter how hard you try, some persons may not approve or may not care much whether or not you approve of what they are doing (unless they are approval seekers themselves). The world is not fair! Approval seekers stay angry because they are not getting what they want. Sometimes they are not sure what they want because they have focused on what others want them to do. The problem with being an approval seeker and also angry is that you can't express this feeling: other people won't approve.

SELF-GLORIFICATION

Self-glorification is what some people call false pride. Those people depending on self-glorification will use any excuse to feel better than others. They put a great deal of energy into covering their own weaknesses. However, they are very interested in the weaknesses of others. Their philosophy of the personal worth is that in order for someone to gain in worth, someone else must lose in worth—a sum zero philosophy: that is, the sum of all human worth is zero. If I see your weakness before you see mine, I go up in worth and you go down. If your football team wins, beating mine, I go down and you go up. If John is a better golfer than Bob, then he is a better person than Bob. If John is a better golfer than anyone else at the country club, he deserves special treatment, concessions that no one else gets. The feelings generated through self-glorification are exhilarating and quick acting. However, they are lost quickly and easily when someone else does better than you or your weakness is revealed.

SELF-ESTEEM

Some people use words like self-worth, self-esteem, self-respect, self-confidence, self-pride, self-assurance, and self-sufficiency interchangeably. But I have assigned different meanings to the words. I use "self-esteem" to represent skills and accomplishments, including both task accomplishment and social skills. If you don't have at least one skill in accomplishing something, you will be low in self-esteem. If you cannot communicate with others well enough to get your social needs met, you will have low self-esteem. So self-esteem involves the ability to get things done and relate to others. The more skills you have (work related, hobbies, or whatever), the higher your self-esteem; the better your communication skills, the higher your self-esteem.

SELF-RESPECT

Self-respect involves trust in your own judgment to do what is right for you. When you can overcome the impulses that give you temporary pleasure or temporary relief while defeating long term goals, you are building self-respect. Self-discipline is the path to self-respect. When you make a personal commitment and keep it, you are adding to your self-respect. Conversely, if you are in the habit of making a commitment and then failing to follow through, your self-respect will suffer. If you give in to the temptation that brings temporary pleasure or relief, your self respect will be low. If you picture a continuum with character weaknesses at one end and self-discipline at the other, your self-respect increases with movement toward the discipline end.

SELF-CONFIDENCE

One can be confident about many things. You may be sure of yourself relative to a particular skill or to your ability to attract other people. These areas are covered above under the heading of self-esteem. However, some very skilled people as well as popular individuals may not feel generally confident. Those who describe themselves as confident usually have particular skills in problem solving. When you have a problem to solve, you must be able to think, figure out what is going on, make decisions, take actions, and make things change. What may surprise you is that confident people have a history of making mistakes and being able to come back, deal with the mistakes, and solve the problems. If you are very careful and seldom make a mistake, don't take risks, and want to make absolutely sure before you act, you probably aren't generally self-confident. Confident people make mistakes and learn from them. Along with their risk-taking they eventually build up a track record of successes. They develop back-up responses so if their first try does not work they have several other ways to do it. The more back-up responses you have, the more confident you are.

Wisdom can be an added dimension to self-confidence. The discernment to identify those problems which cannot be solved frees energy for those that can. The broad perspective of wisdom allows acceptance of limitations as a fact of our humanness—a fact that need not detract from our problem-solving ability and therefore not diminish our self-confidence.

Developing a Positive Self Concept

A positive picture of yourself is very important and has a "whopper" effect on your life. This good feeling about yourself acts as a buffer in otherwise stressful situations. A mistake made by a person thinking poorly of himself can wound deeply. A similar mistake made by a person with a sense of worth may cause her to chuckle to herself and resolve to do better next time. A low opinion of yourself acts as a magnet for keeping yourself down. The more stressed you are, the more poorly you perform, and the poor performance creates more stress.

For example, Tom was a model of success. He was a football hero throughout high school and college. He was also bright and graduated from college near the top of his class. He returned home to farm with his father and brother. Within a few years he mastered his craft to the extent that his father, who managed the farm, would seek Tom's counsel before making important decisions. At age forty-two Tom had achieved most of the goals he had in mind as a thirty year-old. Life had dealt him a good hand.

Tom sat in a tattered chair in a room he used for his farm office which occupied one end of an equipment barn. Papers, old bills, scratch sheets, and long trails of adding machine paper were scattered on his desk and on the floor. It was winter time and cold. An electric heater warmed his feet. Tom kicked over the heater in anger. The safety switch turned the heater off, but Tom didn't care if the place burned down with him in it. He was trying to prepare his financial statement to present to the bank. It was loan time again. Even with a bumper crop next year, he would still be heavily in debt. Over and over in his mind he replayed the thought, "I can't take care of my family. I'm not a man!"

That night his wife June tried to comfort him by initiating love-making which used to cure all of his ills. This time he was impotent! Now he knew he was something less than a real man. He was worthless. He felt weak. He couldn't face the lenders that day. Suicide crossed his mind.

Tom had moved from a highly positive self concept to a totally worthless self perception. Tom had always been at risk because of the manner in which he originally determined his own worth. On three of the six parts of self appraisal—self-esteem, self-confidence, self-glorification, he scored very high. The other three he had thought very little about. Tom was very high in self-esteem (as defined above). He was intelligent and skilled at many things. He was a mover and could get things done. Tom was also a good communicator and had many friends and admirers, of both sexes and all ages. He was certainly confident. He could solve problems, both with people and in business. What gave Tom his biggest high was self-glorification. He was better than his friends at most things, from bowhunting to golf. Although he didn't boast, he felt that he was a better person than most. The importance of Tom's pride was easy to see, because he was a "poor loser." He even got irritable when his old football team hit a losing streak. Losing anything, anywhere, anytime, involving anything related to himself was unacceptable to Tom. He was a winner in life and proud of it. He was special!

When times were better, Tom felt as if he had plenty of self-respect. However, he would be feeling pride rather than self-respect as defined in this book. He was certainly not a disciplined person. He could not trust himself to act in his own best interests. He had been involved in one affair and had enjoyed a few one-night-stands. The last three years he had put on twenty pounds which he swore he would lose each summer or at the beginning of the new year or before he bought his next new suit. He smoked about two packs of cigarettes a day, a habit that worried his children. He had promised to quit but had not done so.

Tom had a great deal of trouble understanding self-love. He knew that he was supposed to love and be loved by others—but love *yourself?* That sounded hokey to him. There must be something wrong with it. The idea of loving yourself without earning it, when things were falling apart, seemed ridiculous to him.

Tom didn't actually depend on direct approval from others, although he got a great deal of it. If he faced disapproval, it didn't worry him much. He generally felt the disapprovers were wrong or perhaps jealous. His accomplishments and pride protected him. So Tom was left to depend on self-esteem, self-glorification (the least stable of the self appraisals), and self-confidence. Self-glorification is the quickest lost when things go wrong. You can lose it in a second. If you have depended on this kind of pride, as Tom had, as a major source of positive feelings, the impact of losing it is heavily felt. Depression may follow, which negatively affects both self-esteem and self-confidence. You don't feel like doing much. As stated earlier, self-confidence is better maintained with wisdom to know when you are "beating a dead horse." Pride kept Tom from developing wisdom. He could not give up—even when facing the impossible. Therefore, in time his confidence dwindled.

Tom was forced by his wife, with the help of his family doctor, to see a psychologist. His pride told him this was just another show of weakness. After the second visit, Tom's homework was each day to make one small, personal commitment related to his well being. His first commitment was to get out of bed by 7:00 A.M. He had been sleeping late since he became depressed. Getting up earlier was a bit difficult but managable. The second day he agreed to stop complaining to his family. The third day he decided to visit his mother whom he had been avoiding. By the beginning of the fourth day Tom already was feeling better about himself. He was developing a sense of trust for himself—self-respect.

In the next several weeks Tom was coached by the psychologist to love himself. This was a much harder task for him. However, the result was worth the effort.

He learned the strength of humility. Humility is a freedom from pride. Tom was able to face his weaknesses, his human frailty, and still maintain a sense of worth. He truly loved himself, not for what he should have been, but for what he was.

PRIDE

Pride is most often a mixture of self-esteem and self-glorification. Pride is much more than being pleased with an act; pride attaches your worth as a person to that act. So to lose pride is to lose self. Pride can get the most out of you—get you to go beyond your limits. It can cause a marine to charge into an enemy machine gun. Pride is a tremendous resource when it propels you to victory, but pride is a destroyer when lost because your psychological self is part of that loss. No human being can completely free himself from pride, but you can learn to decrease the degree to which you depend on it. The more you depend on pride for your feelings of self-worth, the greater your risk of total loss of self when times are tough.

Exercises

1. Building Self-Respect: This is always a good place to start because the process is simple and usually successful in a short time. As described in Tom's story, you start by making personal commitments related to your own well-being. Avoid making commitments that are too difficult or too great. Many times individuals commit to the things they want most and are least likely to

honor. To quit smoking or lose ten pounds or run three miles per day at the beginning of an exercise program are a few of these over-commitments—all good goals but not the place to start. The first commitments that you make must be ones that you WILL keep.

2. Building Self-Love: Find a time to be alone. Arrange the room so that there are six (or more, if you desire) places for others to sit, arranged in a circle. Imagine that loving persons you have selected from your past or present will each take a seat in your group. Your group members may be family, friends, teachers, even persons who are deceased. The criteria for choosing a person is that if that person were actually present, he or she would have loving things to say to you.

Next visualize each member of your group sitting in a particular chair in the circle. Then "listen" as you have each in turn "say" loving things to you. Imagine the voices coming from the place assigned to each particular person. Don't rush: allow as long as it would actually take for each person to express his feelings. Your part is to let yourself feel something emotional not just *think it* in your head but *feel* it in the midsection of your body.

After you have had all the members of your group speak, you now express your feelings *out loud* to each group member. The content doesn't have to be loving—just whatever you would want to express to them if they were there.

Next place two of the chairs opposite each other. Sit in one chair and imagine that another image of yourself is sitting across from you. Say some loving things to yourself. Now move into the opposite chair and accept these expressions of love on an emotional, or feeling, level. The exercises may sound "flakey" to you, and that's all right. Do them anyway. Take my word for it, they are important. If you don't feel anything, keep trying. Imagine God loving you and your accepting this love. This must affect your feelings and be more than just a thought.

3. Reducing Dependence on Self-Glorification: Self-glorification is the most risky way to feel positive about yourself, the one most open to distortion and the one most easily and quickly lost. For these reasons it is wise to keep your dependence on self-glorification to a minimum. If you are to avoid the pain of a sudden loss of pride, you must also avoid the excitement of glorifying your victories. A dose of humility will balance your tendency to exaggerate your importance. Frederick Buechner, a minister and philosopher, says of humility, "The true humility doesn't consist of thinking ill of yourself but of not thinking of yourself much differently from the way you'd be apt to think of anybody else. It is the capacity for being no more and no less pleased when you play your own hand well than when your opponents do."

The key is to be realistic. If your bowling team wins, remind yourself that it was a team effort. If your crop yields more than your neighbor, thank God for his help. Take a long look at your own weaknesses. Do a moral inventory of your life—in writing. This is the fourth of the twelfth steps followed by Alcoholics Anonymous. Doing this will take away the need to put up a front. "What a killjoy!," you may be thinking. I know this strategy will tone down some of your potential pleasure in life. That's okay: you will be motivated to make up for the lost good feelings about yourself in healthier and more substantial ways.

CHAPTER 5

Coping With Conflict

The farm crisis is fraught with conflict. The conflict may be within yourself, with your family, with lenders, with the government, with neighbors, and perhaps with God. The function of conflict is to motivate you when you think you might lose something. The behaviors of this conflict stage are designed to keep you from losing. You are already familiar with victim behaviors presented in chapter 2; these are sometimes part of the conflict experience as well. Chapter 2 also outlined two other ways of coping with conflict: redefining and problem solving. Chapter 3 developed the options open to you after you lose the conflict. The present chapter will focus on handling conflict when another person is trying to make a victim out of you and on understanding conflict in families. Included also are some facts about "mediation," a new method used in some states to ease tensions between farmers and lenders.

Coping With Victims

One characteristic of victims is that they want others to feel bad with them. A farm wife whose husband feels victimized by the weather may experience an unprovoked attack by him or an invitation for her to also feel victimized by the weather. (The intent of the invitation will probably be subtle and unknown even to the husband.) It is a fact that when we human beings feel bad, we derive some comfort from having those around us also feel bad. Conversely, when we are feeling victimized, we may be offended by those around us who maintain high spirits. Since this reaction of victims is natural, you can probably understand why a victim invites you to feel badly, but all of us have a right to avoid these uncomfortable emotions.

You may remember from chapter two that victim behaviors are manipulations in which one person seeks to dominate and control the emotions of another. To manipulate, the conniver must be able to predict the response of the person being manipulated. The rule of thumb for blocking manipulation is to *do the unexpected.* In the situation above the husband, Milburn, listens to an extended weather forecast of rain for several days: no relief in sight. He turns on Mary who has been doing double-duty since Milburn let some of the farm hands go. "We live in a pigpen! You used to be a decent housekeeper but lately you haven't done anything well."

What Milburn expects is for Mary to get defensive, make excuses, attack him back or to feel guilty because she hasn't pleased him. If she makes excuses or attacks, he will have a good fight on his hands. He can't fight the weather, but he can displace his aroused anger on Mary. If she feels guilty he will have a fellow sufferer in the house—someone else to take part of the blame for his pain.

Mary could tell him off. She could really get him where it hurts and tell him that if he were a better manager, she would only have her usual chores, which she could easily handle. However, she need not take the victim role and feel a need to strike back. If she sees the reality of the situation, she knows Milburn is not mad at her. He is speaking from his own pain. He might also be short-tempered if he had a bad toothache or a high fever. She speaks to his pain—not to his attack. "Milburn, I know you have had a tough day. I heard the bad weather report also. Sit down and let me get you a cup of coffee."

Milburn may continue to argue. Mary can nurture as long as he can complain. Eventually she will prevail because his irrational behavior will become obvious to him and he will stop.

The point is that, without your cooperation, nobody can control your emotions and keep you in a victim position. If someone else wants to make you feel bad by dominating you emotionally, it is as if he jabs a knife one inch from your skin. If you don't cooperate, no harm is done. If you allow yourself to be emotionally controlled, it is as if you lunge forward and impale your own body and then say, "You s.o.b., you stabbed me." Attempts at emotional domination are usually obvious. It is surprising how many of us choose out of habit to cooperate in our own pain. There are other choices.

Julia, Sally, and Hazel are all farm wives in an economic crisis. To help out, each has gone to work at the local shirt factory. Their supervisor has worked in the factory all her adult life. She is a bitter woman who never married. She remembers jealously her three employees when times were better. They seemed so high and mighty. This morning she greets them with, "Glad to see you all deciding to work for a living instead of living off the government." Julia responds with embarrassment; Sally walks off the job in anger; but Hazel thinks, "I have a problem. I need to figure out how to enjoy this job in spite of the fact I have a supervisor who takes pleasure in ridiculing me." There is no doubt which of these women suffered after the incident.

So you have choices. A good measure of maturity is the ability to exercise different choices in similar situations. Lower animals don't have this luxury. They are compelled by instinct to do a particular thing in a particular situation. The behavior dictated by instinct may or may not work. Confident people have developed backup responses so that when their first response fails, they act a second way and a third and so on until they accomplish their goals. You can develop alternative behaviors and also alternative ways of defining situations. *In a conflict situation you can choose to redefine or rethink a threatening situation as no threat at all.*

Redefining Threats

For an event to be a real threat it must signal a loss to you that will be followed by physical or emotional pain. Human beings are symbolic creatures, so many of our threats are symbolic rather than actual. Symbols can be interpreted in many ways. For example, *the need to be right* can be a serious symbolic loss if it is not satisfied. Ned was deciding whom to vote for in the coming election for state representative. Both candidates courted the farm vote. Both were moderate in their political histories, and both were trying to appear a bit more conservative in this election. Ned ultimately decided to support the candidate who was older than the other by five years, believing he should have learned something during those five years. At the moment he chose, the candidate took on a new characteristic: he became *his* candidate. Now attacks on the candidate are personal attacks on Ned himself. It appeared to him that his neighbor, who pointed out a flaw in his candidate, was saying Ned was stupid for making a bad choice. Ned argued strongly for the virtue of a man that only a few days earlier he had chosen on a fifty-one to forty-nine percent basis.

You learned to value "being right" in early childhood. At that time you were so dependent on your parents that you had to "prove" to them that you were right in wanting whatever you wanted or you wouldn't get it. If other children could prove you wrong, you might be teased and feel like a loser. Being right was all-important in childhood, but the need to be right can be a weakness in the adult. You don't need to fear "looking bad" when someone has a better argument. It's okay to argue your point in a disagreement that is current; however, if the discussion is about a past right or wrong, it is a waste of time. It will only lead to blame and, remember, blame is the most damaging of the "victim behaviors"—damaging to the blam*er*, that is. Being right is part of self-glorification: to be right is to feel better than those with different opinions. Remember that self-glorification is an illusion. Greg, in the previous chapter, made a poor judgment on the land purchase. He lost money. If he feels he has to be right, he will also lose self-worth—unnecessarily. If he rethinks the situation and realizes that he doesn't have to live in the illusion of being perfect, he can remind himself that mistakes are human. Now he still loses the money but does not lose "self." When you don't have to "be right" to feel safe, you avoid a great deal of conflict.

Another illusion is that other people are responsible

for pleasing you. John was counting on two friends coming to his house to watch the Superbowl. Both cancelled, and John was upset. Because his friends had decided to spend their day doing something else (which they had every right to do), he didn't enjoy the game much. His loss seemed deeper than his friends' absence. If John redefines his loss as only companionship and recognizes that his friends are not responsible to please him, he is free to enjoy the game—that is, to please himself. He can allow his friends the freedom to do what they want without destroying his relationship with them.

Conflict Resolution (Solving problems with others)

Redefining involves changing something in your mind to lessen or wipe out the threat of loss. Sometimes the threat is real and the potential loss is significant. Conflicts that may result in loss can be resolved so there is no loss. If the conflict is with another person with whom you will have a continuing relationship, it is important that all those involved are satisfied with the resolution. The farm financial crisis produces conflict with family, neighbors, local businessmen and government agencies. Many of the individuals involved will have a continuing relationship—good or bad.

To resolve conflict without destroying the relationships, you need to develop skills in debating your own interests, in listening to the other side and in compromising. Debating your interests doesn't mean defending your original position. Think about what you really *need* to avoid feeling a major loss. For example, Harvey was in conflict with his wife about her returning to a teaching career. He said it made him feel less like a man when his wife had to work, yet the family needed her income. As the couple talked, it became clear that the real need of Harvey was reassurance that his wife was not planning to leave him. He feared the job was only the first step on her way out. So *you must know what you really need and be able to talk about it in a rational manner.*

In any communication the real power is with the listener. No matter how well a thing is explained, if the listener doesn't hear it, there is no communication. Listening means that you pay full attention to what the other person is saying, not thinking of what you are going to say next. The goal of the listener is also important. You can listen for faults in what is being said, or you can listen for a possible compromise or solution. The major reasons for listening to a competitor in a conflict situation is to gather facts, ideas, and feelings related to solving your problems with the others involved.

Compromise means that all parties involved will have to give up something, and that no one person will have to lose it all. It is easy for most readers to think in terms of win/lose resolution to conflict. One person gets what he has been competing for, and the other gets nothing. However, compromise—a fair solution for all—takes creative thought. Both sides need to have as their goal a creative solution to the conflict instead of winning the conflict.

Conflict resolution with lenders and neighbors is important but even more important is resolving family conflicts—particularly between spouses. Living together successfully as married people requires mutual respect, mutual trust, a willingness to confront problems, a willingness to compromise, and a willingness to forgive. Love alone will not keep a family together.

Respect depends upon maintaining a high level of regard for your spouse or other family members. This regard involves being able to understand and accept your mate's differences, to forgive past transgressions, and to focus on cooperation. Respect demands that each person be responsible for his or her behavior and feelings. People who respect each other share work loads with an agreed-upon division of labor. Although healthy individuals may alternate between dependency and responsibility, chronic dependency or chronic control destroys respect. A key to maintaining respect is the ability to solve problems within the relationship without blame.

Trust involves an expectation of the behavior of your marital partner. Trust is absolutely necessary for intimacy to occur. When you open yourself up to your mate, you can be hurt and you won't do it if you don't trust. So marital trust has to do with your feeling of safety in the company of your husband or wife. If you don't trust, you can be expected to take the necessary steps to protect yourself. So, a lack of trust may result in feeling justified in controlling your spouse, which violates your mates sense of safety—then everyone is defensive—and the mistrust grows.

Confronting problems involves a willingness of mates to discuss problems within their relationship. This activity may be the most difficult part of relating. Usually telling someone you care about what you like and don't like about his behavior is potentially unpleasant and requires a trust of your mates's willingness to solve problems. Healthy confrontation includes expressing your own needs and wants and asking for changes in your mate's behavior. These confrontations are successful when all parties refuse to withdraw when feelings are aroused. The confrontation then continues until a workable compromise is reached.

Compromise necessitates putting aside the temporary relief one gets from power and control or from giving in and giving up. The two individuals listen for solutions instead of finding fault. (Remember comments about listening discussed above.) Win/lose arguments are avoided and ultimatums are withheld. For successful compromise, it is better to focus on the problem causing the discomfort before taking a firm stand on what you consider the solutions to the problem; that is, be willing to negotiate and listen to your mate with an open mind rather than push your beliefs about how to handle the situation.

Forgiveness involves putting the incident and negative feelings behind you and starting fresh. Review chapter three for an understanding of turning resentment into grief. Forgiveness needs to be unconditional. Those who say, "I can forgive, but I won't forget," are not forgiving. You do not have to wipe an unpleasant incident out of your conscious mind to forgive, but you do have to put it behind you. A truly forgiven act from the past will have no effect on future relationships. Buechner states, "For both parties, forgiveness means the freedom again to be at peace inside their own skins and to be glad in each other's presence."

Exercises

1. Categorize your recent responses to conflict into victim, redefining or problem solving.
2. Now be a creative thinker and see how many of those situations in the victim category can be moved into one of the other two categories.
3. *Think* about your future conflicts and *decide* where you want to be. If you decide to think and solve problems, the following steps should be included:

(a) Be willing to change yourself *first.* In my marriage counseling I often find two people willing to change *second.*

(b) Decide on the outcome that will solve your problem. There are usually many possible outcomes which will contribute to the relief of your discomfort. If the outcome is to be useful, it should lead to an action. If you can't come up with an action, come up with a *new outcome.*

(c) Think of as many actions as you can. Pick one and try it out.

(d) Evaluate your results. If you made progress toward achieving your outcome—great! If you didn't, take another action. Keep it up until the problem is solved.

Farm Finance Mediation

In 1986 the Minnesota Legislature instituted mandatory farm credit mediation, a method of resolving conflicts between farmers and financial institutions. Because of this legislation no creditor can start a proceeding to collect a debt against agricultural property without first attempting mediation. Other states have followed. Mississippi passed a bill in the 1987 legislature for voluntary mediation. *Mediation* employs a neutral mediator who is there to see that fairness is achieved. "Fairness" means that all parties involved agree that the solutions are fair. The mediator follows a structured method, which is very much different from the usual judicial procedure. Our courts operate in an adversarial atmosphere designed to produce a winner and a loser. The focus is on the history of the problem—what happened, who was at fault—and law, when the facts are known the consequences are spelled out.

Mediation seeks agreement on the facts, agreement on the problems to be solved and the creation of options to solve the problems. Any solution on which all parties agree is acceptable, giving a great deal of flexibility. In order to accomplish this task an atmosphere of safety must be developed among the members of the problem solving group which would normally be competitive with each other. The mediators must agree to act "in good faith"—that is, do their best to reach a solution. Cheap-shots, foot-dragging, negativism, and guardedness are counter productive and get the procedure into an adversarial atmosphere. Good faith bargainers attend all meetings and are willing to consider alternatives to the original terms of the debt. A second requirement for a feeling of safety is for the participants to agree on confidentiality; that is, matters discussed in the mediation are designed to explore creative options and cannot be used later in court.

Your part in the mediation process is to act in "good faith" as outlined above. Realistically, you will be anxious and that's okay; so will the lenders on the other side. Nobody likes conflict very much. However, you both have a mediator trained to protect all involved. If the process of mediation is to work, you first have to trust that process. That is a risk for you. The alternative is even more risky. You are more likely to get a square deal in mediation than in the courts. You also have the right to have your own attorney review any agreement before you sign it.

In the Short Rows:

A Final Note from Neill

It is our hope that, having read this far, you now understand the farmers' plight a little better and that you may be more prepared to deal with the problems brought on by this crisis. There is one other factor I feel we need to touch on before putting the chairs in the wagon, though.

That is the Personal Touch: caring, love, friendship, Christian concern—whatever you want to call it. Unless you live in the middle of a metropolis, you undoubtedly know someone who is in some stage of going through this type of economic crisis. Even if you are going through it yourself, you may still be the key to breaking the pattern of despair, depression, and isolation in another person's life. Reach out to them.

Everybody needs somebody. David had Jonathan; Ruth had Naomi; Christ had Peter, James, and John. (And if you think their going to sleep during prayertime was bad, just sit in the choir next Sunday and count the deacons who sleep during the sermon!) Yours may be the shoulder someone else needs to lean on, or even cry on. Be there.

If I was to dedicate this book, I would do it to Dude and Beau, brothers and friends, of whom I could say, like Shadrach, Meshach, and Abednego:

"We been through the fire together!"